# Cognitive Science: Philosophy of Mind

# Cognitive Science: Philosophy of Mind

Edited by Ernest Simmonds

CLANRYE
INTERNATIONAL
www.clanryeinternational.com

Clanrye International,
750 Third Avenue, 9th Floor,
New York, NY 10017, USA



ISBN: 978-1-63240-723-8

**Cataloging-in-Publication Data**

Cognitive science : philosophy of mind / edited by Ernest Simmonds.
p. cm.
Includes bibliographical references and index.
ISBN 978-1-63240-723-8
1. Cognitive science. 2. Philosophy of mind. 3. Cognition. 4. Cognitive psychology.
I. Simmonds, Ernest.
BF311 .C64 2018
153--dc23

For information on all Clanrye International publications
visit our website at www.clanryeinternational.com

CLANRYE INTERNATIONAL

Printed in China.

# Contents

# Preface

Cognitive science refers to the study and analysis of the various processes which a mind performs. It examines the behavioural and intelligence characteristics present in the brain. Mental processes like reasoning, emotions, logic, memory, attention, language, etc. are studied under this area. This book aims to shed light on some of the unexplored aspects of cognitive science. The topics covered in this extensive textbook deal with the core aspects of the field. It is an essential guide for both academicians and those who wish to pursue this discipline further.

To facilitate a deeper understanding of the contents of this book a short introduction of every chapter is written below:

Chapter 1- This chapter focuses on the problems and issues faced by philosophy in today's time. Dualism, materialism, mind-body problem, global justice, etc. are some of the topics intensively discussed. This section will provide an integrated understanding of the contemporary issues related to philosophy.

Chapter 2- The process of acquiring and gathering knowledge is known as cognition. It incorporates processes like memory, learning, attention, decision-making, etc. Some of the common experiments in the field of cognition are serial position, Brown-Peterson, memory span, visual search and word superiority. This chapter is an overview of the subject matter incorporating all the major aspects of cognition.

Chapter 3- Philosophy of artificial intelligence questions the progress of artificial intelligence. It questions the advances in artificial intelligence and its ability to gain consciousness and develop a mind similar to humans. This section provides a plethora of interdisciplinary topics for better comprehension of artificial intelligence.

Chapter 4- The non-computational theories studied under this chapter are theory of mind and Cartesian theory. These theories focus on the ability of the mind to assign mental states to other persons. This chapter discusses non-computational theories in a critical manner providing key analysis to the subject matter.

Chapter 5- The mind cannot be explained in only neurological or objective terms. The following section studies the subjective nature of the mind. It focuses on questions which are very fundamental in nature. This section helps the readers in developing a better understanding of the subjective aspect of mind.

I would like to share the credit of this book with my editorial team who worked tirelessly on this book. I owe the completion of this book to the never-ending support of my family, who supported me throughout the project.

**Editor**

1

# Contemporary Issues in Philosophy of Mind

This chapter focuses on the problems and issues faced by philosophy in today's time. Dualism, materialism, mind-body problem, global justice, etc. are some of the topics intensively discussed. This section will provide an integrated understanding of the contemporary issues related to philosophy.

## Contemporary Issues

The contemporary issues in philosophy of mind and cognition delve into several philosophical issues centered around the notion of consciousness. The mind and the cognition are inner realities of the conscious mental life. But sometime these concepts are theorized differently. For instance, the analysis of the concept of cognition has a physical connotation, rather than having strong connection with consciousness. In other words, cognition can be explained not necessarily with reference consciousness. On the other hand, the mind has a metaphysical connotation and often being stated at par with consciousness. The mind is necessarily consciousness. In this course, we will be more dealing with the notion of mind and consciousness, sometimes taking them as synonyms, rather than talking about cognition. Only when we talk about functionalism and connection with reference to the philosophical study of Artificial Intelligence, we would deal with the notion of cognition. So the mind and cognition as two different concepts and have different meaning is indeed a recent phenomenon. We would not seriously get into this debate here, rather would like to deal with the notion of consciousness or the mind as a unitary mental phenomenon.

The concept of mind has been theorized in many ways, the issues here are concerned with different characterization of the mind. For instance, the mind has been traditionally conceived as a transcendental reality. To conceptualize its existence, we need to transcend the very mode in which we have knowledge about the external world. The objects in the external world exist independent of the mind. The knowledge of the external world and the knowledge of the mind are different in the sense that they are conceived differently. More precisely, the consciousness transcends the epistemic mode in which it conceives things. In this regard, the issues that are discussed are the transcendent mind, dualism, mind-body identity theory, reductionism, subjectivity, qualia, meaning, creativity, freewill and personal Identity. As philosophers have been debating and discussing about the nature of the mind from the time immemorial, these issues are found significant with reference to the central question namely: what is consciousness? To answer this question, there are two things we need to focus on in our study, first the analysis of this question, and second is the nature of the philosophical inquiry.

The question, 'what is consciousness?' is indeed a complex one? As Paul Johnston, in the context of responding to Wittgenstein's statement 'consciousness is life itself' says we need to study the concept of consciousness and its relationship with other concepts. It is interesting to note that large numbers of philosophers still consider that consciousness is something beyond the ontology of the physical. And this is happening in 21st century when we notice significant progress has been made in the field of neuroscience, artificial intelligence, cognitive psychology, and many other scientific disciplines regarding the brain studies are making their claim very straight that 'consciousness is something to be scientifically studied.' Meaning thereby they do not see consciousness as something special and exist beyond the neurophysiology of the brain. Hence, the ontology of consciousness depends on the ontology of the brain. Hence, we begin with this issue that 'what is the ontology of consciousness?

Our second focal point is to talk about the nature of philosophical inquiry. In other words, what is the philosophical inquiry all about the concept of consciousness/ mind? Philosophy is a rational enterprise that studies the problem without making a dogmatic or nihilistic claim. Its rational approach is shown in the usage of concepts in a *perspective* and evaluates them *critically*. A perspective signifies the theoretical standpoint from which a particular concept is viewed, whereas the critical evaluation refers to the analysis of the concepts used in the theory. Philosophy aims at clarifying concepts, so that the problem associated with the study is resolved. This has been the basic approach of analytic philosophy. We will be using linguistic conceptual analysis as method to discuss the issues that are stated above. As some of the analytic philosophers of mind are influenced by the scientific study of consciousness, our study will focus on those theories to capture the nuances of the concept of mind.

The whole endeavour here is to understand the issues from theoretical perspective of linguistic philosophy. As we all know that philosophy is a foundational discipline investigates things logically, rather than making the study completely empirical. Science does pursue empirical study – it studies the empirical facts. In fact, science tries to explain the origin of these empirical facts. In this context, Wittgenstein gives a fitting remark in the Tractatus. He writes, "Philosophy is not one of the natural sciences. (The word 'philosophy' must mean something whose place is above or below the natural sciences, not beside them)." Philosophers do study 'facts' emphasizing clarity and elucidation. Science, on the other hand, focuses on verification and testing. Philosophy does not show interest to subject itself to testability. Philosophical knowledge is conceptual. And, philosophical explanations are about understanding the nature and use of the concept in details. Of course these concepts are related to life. Hence, demands clarification and elucidation.

Philosophy is a humanistic discipline. It studies 'the ultimate and the general features of all existence'. In other words, how genuinely we encounter them in our everyday life "we think about the existence of the world and the destiny of man in the world." There is no single method or single mode of doing philosophy. Philosophical activities is carried in multiple ways, thus it is inevitable that philosophers are bound to have Pluralism and Disagreement. Philosophy presents a pluralistic scenario of theories, thesis and methods. "Disagreement is the permanent facts of the philosophical solution. Philosophy grows in best way when there is disagreement. This also reveals the inner dynamics of philosophy as an ascientific theoretical discipline. Disagreement generates possibility of new philosophy with a new method and new solution." So, the progress in philosophy depends on not merely rejection and confirmation of philosophical ideas, rather it is more about acceptance. And, the acceptance of a philosophical knowledge demands rational consensus. It is because philosophical worldview does not overrule one another. Rather proliferates in many ways pertaining to our everyday activities in life. For example, within analytic naturalism, there are many theories of mind, such as biological naturalism, functionalism, physicalism, emergentism, etc. However, their quest for truth lies in their engagement and description of the world. Secondly, these theories do not discover new facts, rather present new knowledge of facts. Philosophizing, thus, needs critical thinking and concept formation. Not merely with justification, but reflection. It is because in the process of articulating new concept the old concept is displaced. According to Pradhan, the concept displacement is very much part of the philosophical inquiry. Philosophical knowledge concerning progress is ought to be intelligible, and that would help building rational consensus.

## Transcendental Consciousness

The notion of consciousness is delved from the transcendental perspective, in the *Upanisads* and in other classical philosophical texts. Consciousness is transcendental which we find significantly raised in the Upanisads, particularly with reference to the questions on – what is the nature of the self/ ātman/ soul? This question is not typical to the Indian philosophical thinking, rather is also found in the Western tradition as well. For instance, the classical Greek philosophers were also concerned about knowing the nature of the self/ soul? We begin our discussion with the notion of *transcendental mind.* The mind as a transcendental reality is studied in the *Upanishads* and also in the Greek Philosophy of Plato and Aristotle. The transcendental mind is a metaphysical thesis that maintains the mind or the soul is an immaterial substance. The mind is neither identical with the body nor is caused by the bodily organism. It is independent of the function of the body. The body is conceptualized as a material entity whereas the mind is construed as spiritual.

The *transcendental* theorization of the mind maintains that 'no mind; there is no matter,'[1] writes Robinson. It means there is no physical existence of the reality conceived beyond the presence of the mind. The mind is real and transcends the given physical existence of the reality. To say that the mind is real is to begin with ontological status of existence of the mind. The study of the ontology as initiated by the philosophers is based on the questions: 'Is there a mind?', 'What kind of stuff it is?', and 'What is the nature of mental life? These questions are delved philosophically broadly from point of view of metaphysics, epistemology, and ethics. As metaphysics was central to the philosophical discourse traditionally it directly dealt with the ontology of the mind. To assert the metaphysical position that mind is real is to claim that consciousness is the underlying principle essential for the explanation of the reality as a whole and as such. The mind is understood in the form of nonphysical conscious being on which everything in the reality are grounded. In this regard, Robinson's point is clear that if the ontology of consciousness is not presupposed then no explanation of the reality as a whole is possible. Thus, the dictum 'no mind there is no matter' is meaningful. On the contrary, there are also metaphysical positions which enchant the ontology of the matter. In other words, there is also a metaphysical position which uphold to the thesis that 'no matter; never mind.' The primacy is given to the existence of matter as the first principle to explain the very foundation of the reality. The materialists maintain this theoretical position strongly. So, materialism as a metaphysical thesis has tremendous impact of the taxonomy of the mind. The spiritualists, on the other hand, maintain that the existence of the mind/ consciousness is primordial – the life and its varied forms and structures have emerged from consciousness. Thus, immaterialism or the monistic spiritualism holds the maxim 'no mind; there is no matter,' goes well to justify that consciousness is the metaphysical form of the reality.

The philosophical debate regarding the ontological status of the *mind* and the *matter* is indeed primitive. But this theoretical position is resurrected in the modern period especially by Rene Descartes. The ontological or substance dualism that follows from Descartes' investigation shows the mind/consciousness is as fundamental as the matter/ extension. The mind and the matter exist and are ontologically real. Thus, follows dualism as a metaphysical thesis that theorizes the existing material bodies in the world and the existing mental phenomena beyond the realm of the physical. Dualism as a taxonomy of the mind shows the scope of continuing the spiritualistic approach to the study of the mind when the modern science was emerging vigorously as fundamental discipline for the explanation of the reality. Hence, there are evidences that the physical in its ontological form is not undermined either. And, the mind as substance gets a special status from the *epistemological* perspective. The ontology of the mind is established from through epistemology. In this way, Descartes opens up a new debate pertaining to the very nature of the mind. And, it has been a major source for the contemporary discourses in philosophy of mind and cognition. In the contemporary period, when the success of science and technology is enchanted over and over again monistic materialism as taxonomy of the mind eventually has been taking over. The mind is delved from the perspective of many special sciences, such as, neuroscience, psychology, and artificial intelligence. Rather the scientific study of the mind has given birth to a new and a special science of mind known as *cognitive science.* The cognitive science systematically studies the nature of human cognition and cognitive processes in order to reveal the very mystery that is surrounded around the concept of mind or cognition. So, the contemporary debate has yet to give up the ontological questions, rather throws serious reflection on the very developments that are happening within the paradigm of cognitive science. The eliminative materialism as taxonomy of the mind shows the success of the paradigm shift that has happened in the very discourse of the study of the mind. For the eliminative materialists, the mind is no more a mystical and a metaphysical reality, rather it is an *epiphenomenon.* The denial of the ontological status has not wiped out the notion of mind *per se* from the philosophical study, rather has shown that there many aspects, as such human *subjectivity* – the *first person point of view* of experience of the reality is an undeniable fact. Unless the subjectivity is retained many epistemological and ethical questions pertaining to knowledge and value respectively will remain unanswered which will further pose serious threat to the *ontological* status of humanity and *human institutions.*

However, 'the concept of mind as transcendent reality is indeed venerable', in the sense that it has more of spiritual relevance. As if it only satisfies the query of the spiritual person and in philosophy it stays close because of its metaphysical resilience. In Indian Philosophy, at least some of the major philosophical systems, such as, *Sānkhya*, *Vedānta*, and *Nyāya* which accept the authority of the *Vedas* and the *Upanisads* maintains the existence Brahman as substance has metaphysical relevance. The *Brahman* is the being conceptualized as the creator of the whole universe. Hence could be treated at par with the theological concept of the God. As a *substance* – metaphysical principle does not depend upon nor reducible to the matter. The *Brahman* as a *pure consciousness* is ontologically distinct from the existence of the basic form of the matter. The individual consciousness/ the mind are due to the presence of the soul or the ātman which is a part of *pure consciousness* - universal consciousness. The Greek philosophical tradition also gives importance to the ontological status of the mind. The entire metaphysical discourse in the pre-Socratic period and in the Socratic period does not diminish the significance of the mind. Pythagoras, Plato, and Aristotle have given special consideration in their philosophical discourses. Robinson writes, "The mind is a thing apart; neither dependent on nor reducible to matter; not sharing matter's features or fate." For Pythagoras, the mathematical knowledge is comprehended by the virtue of mind's rational character. Reason has a special role in forming judgment. To understand how reason operates is a special kind of self-discovery. In other words, knowing the function of the rational mind is almost in the line of knowing the eternal truth. The knowledge of the true existence of the numbers is less personal, rather is abstract which is like *knowing thyself.* True knowledge, for Pythagoras, is eternal. Plato also maintains that reason plays an important role in construing the very existence of life. We all encounter many illusory facts in our everyday life. Unless we question them *rationally* the very mode existence, it would be difficult to attain true knowledge. Plato's famous parable

– *the parable of cave*[3] explicates the role of reason in understanding truth. The sense perceptual knowledge could be illusory; whereas knowledge construed by reason is significant. True knowledge is transcendental. The soul has a transcendental grounding. It not only relates to the perceptual experiences, but also relates to emotional and intellectual operations of the mind. The soul is divine; its existence can be comprehended through reason. Thus, the ontology of the mind/ soul is based on a metaphysical and epistemological thesis.

The epistemic aspect of the study asserts that there must be knowledge independent of experience. Senses no doubt form experiences, but the existence of the *knowledge of forms* in Plato's philosophy or the *knowledge of numbers* in Pythagoras' philosophy refers to non-sensory knowledge. In other words, knowledge (true knowledge) is being characterized as non-material. The ontology of such knowledge does not derive through sense experience, rather depends upon the function of reason – the rational mind. Since the mind per se is immaterial, the rational feature of the mind is also construed as immaterial. The mind as such being immaterial is 'embodied' without itself being a body. In other words, the mind is not identical with the body. Rather conceived transcending the physical function of the body. Aristotle emphasized the function of the mind with reference to the notion of *efficient cause* -the agent. The mind represents the *thinking agent* which carries the creative power in making something possible. The efficient cause in itself is different from the *material cause*, because it initiates the change. In other words, all kinds of change that occurs in the case of an *embodied being* are not due to the materiality of the body, rather presupposes the existence of *the soul* – the thinking agent. It is analogues to the notion of God who is the creative power behind the sensible universe and the world of nature. The God is the *unmoved mover*. He is being responsible for change but does not get changed. As an agent, for instance, *I* am instrumental in *causing* certain voluntary actions. *I* not only cause them or make them happen, but also own them, in the sense that *I* am responsible for committing these activities. By *causing* these actions, *my* ownership of these actions does not change. Thus, the ontological status of the *agent* as an efficient cause remains significant in opening up the *autonomy of agency* in philosophy of mind and ethics but also in the contemporary discourse of philosophy of action and moral psychology.

The notion of autonomy of agency is attached to the notion of *free-will*. The freedom of will is an essential attribute of *thinking mind*. Freedom expressed in action is derived from the content of thought. Freedom is an experiential fact intrinsic to the thinking mind. A conscious person can perform action freely and rationally. For instance, the person as an autonomous being not only gives orders, but also obeys and disobeys the orders of the others. The freedom of will thus expressed in the very engagement of the *being in the world*. So far as the ability to think and perform action is concerned free-will is not determined by the causal principle that governs the function of the world. The world as a material body represents a system of mechanism which can be causally explained. In this regard, the principle of causality puts a closure to the entire function of the world. Philosophers have been debating whether such a closure can be put to explain human action. That is to say, can action be determined at all? The determinism if retained with rigor then it nullifies the ontology of *free-will*, and thereby the status of human agency. The indeterminists, on the other hand, asserts on the thesis that human mind is *immaterial*.

The *epistemic ability* of the *immaterial mind* is studied in the *Upanisads*. The ātman that represents this *immaterial mind* is itself the source of *knowledge*. It can only *illuminate* itself. The notion of self-knowledge is Indian philosophy not only means *self-consciousness,* but also means realization of one own being, i.e., *self-realization*. The aspiration to have *self-knowledge* as an epistemic ability is very well portrayed in dialogue between Maitṛéyi and Yājñavalka, in the *Bṛihadāranyaka Upanisad*. Maitṛéyi expresses her desire for knowledge which would *enable* her eternally in realizing everything. Understanding Maitṛéyi's inquisitiveness for knowledge, Yājñavalka explains to her the power of *self-knowledge*. The self-knowledge is essential because everything is centered around the notion of self. Hence, "the self is to be seen, to be heard, to be perceived, to be marked, and everything else is only for the sake of the self." The notion of centrality of the self is conceived beyond the conceptual domains in which the self is *karta* (the agent of all kinds of action) as well as *bokta* (the enjoyer of all the results of action). Rather, the self is the *witness* (*sāksin*) of all the mental events and actions. The concept of *sāksin* is illustrated through a parable of two birds in the *Brihadāranyaka Upanisad*.

The *Upanisads* have seriously delved into the nature of the self, particularly its *immaterial* ontological foundation with regard to the *value* of life. For instance, the dialogue between Yama and Nachikéta, in *Katha Upanisad* expedites the *telos* of the immateriality of the soul. Nachiketa's curiosity to learn the truth of life after death highlights once again the whole purpose of knowing the nature of the self which is *immortal* and *immaterial*. Hence the knowledge that enables a person to care for worldly goods is not sufficient for the *self-knowledge* which involves the truth concerning the *spirituality of life*. In this regard, *self-knowledge* emphasizes the inner reflective attitude of the self, which helps in evaluating and understanding the very purpose of living a *meaningful life*. The value of life here is conceptualized beyond the existence of *being* and *non-being*, the material and non-material existence. This conceptualization of self-knowledge aims to arrive at the level of consciousness which is only consciousness. Consciousness transcending all the binaries arrives at a state of *sunyatā*. The concept of *sunyatā* does not mean absence or pure negation. Rather, it is the state of *inexpressible* and *indefinable* presence of self-consciousness. It excludes everything than its own being. The objectivity of itself is hence an indefinable truth. "All factors conceived as essential to individuality are objective constructions and so cannot reveal the self. Whatever is conceivable belongs to the objective world by that every fact to be excluded from the area of consciousness."

## Soul

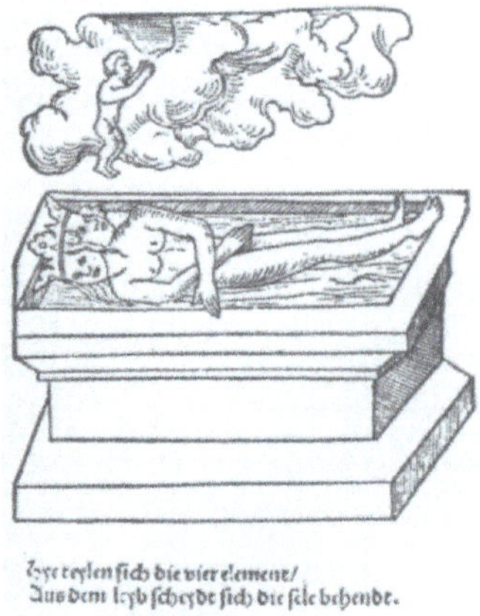


Image of the soul in the Rosarium philosophorum

In many religious, philosophical, and mythological traditions, the soul is the incorporeal essence of a living being. Depending on the philosophical system, a soul can either be mortal or immortal. In Judeo-Christianity, only human beings have immortal souls (although immortality is disputed within Judaism and may have been influenced by Plato). For example, the Catholic theologian Thomas Aquinas attributed "soul" (*anima*) to all organisms but argued that only human souls are immortal. Other religions (most notably Hinduism and Jainism) hold that all biological organisms have souls, as did Aristotle, while some teach that even non-biological entities (such as rivers and mountains) possess souls. The latter belief is called animism.

Greek philosophers, such as Socrates, Plato and Aristotle, understood that the soul must have a logical faculty, the exercise of which was the most divine of human actions. At his defense trial, Socrates even summarized his teaching as nothing other than an exhortation for his fellow Athenians to excel in matters of the psyche since all bodily goods are dependent on such excellence (*Apology* 30a–b).

*Anima mundi* is the concept of a "world soul" connecting all living organisms on planet Earth.

## Philosophical Views

The ancient Greeks used the word "alive" for the concept of being "ensouled", indicating that the earliest surviving western philosophical view believed that the soul was that which gave the body life. The soul was considered the incorporeal or spiritual "breath" that animates (from the Latin, *anima*, cf. "animal") the living organism.

Francis M. Cornford quotes Pindar by saying that the soul sleeps while the limbs are active, but when one is sleeping, the soul is active and reveals "an award of joy or sorrow drawing near" in dreams.

Erwin Rohde writes that an early pre-Pythagorean belief presented the soul as lifeless when it departed the body, and that it retired into Hades with no hope of returning to a body.

## Socrates and Plato

Plato (left) and Aristotle (right), a detail of The School of Athens, a fresco by Raphael

Drawing on the words of his teacher Socrates, Plato considered the psyche to be the essence of a person, being that which decides how we behave. He considered this essence to be an incorporeal, eternal occupant of our being. Socrates says that even after death, the soul exists and is able to think. He believed that as bodies die, the soul is continually reborn in subsequent bodies and Plato believed this as well, however, he thought that only one part of the soul was immortal (*logos*). The Platonic soul consists of three parts:

1. the *logos*, or *logistikon* (mind, nous, or reason)
2. the *thymos*, or *thumetikon* (emotion, spiritedness, or masculine)
3. the *eros*, or *epithumetikon* (appetitive, desire, or feminine)

The parts are located in different regions of the body:

1. *logos* is located in the head, is related to reason and regulates the other part.
2. *thymos* is located near the chest region and is related to anger.
3. *eros* is located in the stomach and is related to one's desires.

Plato also compares the three parts of the soul or psyche to a societal caste system.

According to Plato's theory, the three-part soul is essentially the same thing as a state's class system because, to function well, each part must contribute so that the whole functions well. Logos keeps the other functions of the soul regulated.

## Aristotle

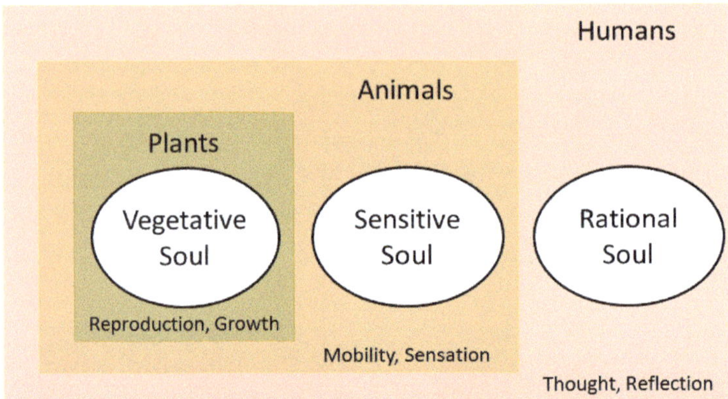


The structure of the souls of plants, animals, and humans, according to Aristotle

Aristotle (384 BC – 322 BC) defined the soul, or *Psūchê*, as the "first actuality" of a naturally organized body, and argued against its separate existence from the physical body. In Aristotle's view, the primary activity, or full actualization, of a living thing constitutes its soul. For example, the full actualization of an eye, as an independent organism, is to see (its purpose or final cause). Another example is that the full actualization of a human being would be living a fully functional human life in accordance with reason (which he considered to be a faculty unique to humanity). For Aristotle, the soul is the organization of the form and matter of a natural being which allows it to strive for its full actualization. This organization between form and matter is necessary for any activity, or functionality, to be possible in a natural being. Using an artifact (non-natural being) as an example, a house is a building for human habituation, but for a house to be actualized requires the material (wood, nails, bricks, etc.) necessary for its actuality (i.e. being a fully functional house). However, this does not imply that a house has a soul. In regards to artifacts, the source of motion that is required for their full actualization is outside of themselves (for example, a builder builds a house). In natural beings, this source of motion is contained within the being itself. Aristotle elaborates on this point when he addresses the faculties of the soul.

The various faculties of the soul, such as nutrition, movement (peculiar to animals), reason (peculiar to humans), sensation (special, common, and incidental) and so forth, when exercised, constitute the "second" actuality, or fulfillment, of the capacity to be alive. For example, someone who falls asleep, as opposed to someone who falls dead, can wake up and live their life, while the latter can no longer do so.

Aristotle identified three hierarchical levels of natural beings: plants, animals, and people. For these groups, he identified three corresponding levels of soul, or biological activity: the nutritive activity of growth, sustenance and reproduction which all life shares; the self-willed motive activity and sensory faculties, which only animals and people have in common; and finally "reason", of which people alone are capable.

Aristotle's discussion of the soul is in his work, *De Anima* (*On the Soul*). Although mostly seen as opposing Plato in regard to the immortality of the soul, a controversy can be found in relation to the fifth chapter of the third book. In this text both interpretations can be argued for, soul as a whole can be deemed mortal and a part called "active intellect" or "active mind" is immortal and eternal. Advocates exist for both sides of the controversy, but it has been understood that there will be permanent disagreement about its final conclusions, as no other Aristotelian text contains this specific point, and this part of *De Anima* is obscure.

## Avicenna and Ibn al-Nafis

Following Aristotle, Avicenna (Ibn Sina) and Ibn al-Nafis, a Persian philosopher, further elaborated upon the Aristotelian understanding of the soul and developed their own theories on the soul. They both made a distinction between the soul and the spirit, and the Avicennian doctrine on the nature of the soul was influential among the Scholastics. Some of Avicenna's views on the soul include the idea that the immortality of the soul is a consequence of its nature, and not a purpose for it to fulfill. In his theory of "The Ten Intellects", he viewed the human soul as the tenth and final intellect.

While he was imprisoned, Avicenna wrote his famous "Floating Man" thought experiment to demonstrate human self-awareness and the substantial nature of the soul. He told his readers to imagine themselves suspended in the air, isolated from all sensations, which includes no sensory contact with even their own bodies. He argues that in this scenario one would still have self-consciousness. He thus concludes that the idea of the self is not logically dependent on any physical thing, and that the soul should not be seen in relative terms, but as a primary given, a substance. This argument was later refined and simplified by René Descartes in epistemic terms, when he stated: "I can abstract from the supposition of all external things, but not from the supposition of my own consciousness."

Avicenna generally supported Aristotle's idea of the soul originating from the heart, whereas Ibn al-Nafis rejected this idea and instead argued that the soul "is related to the entirety and not to one or a few organs". He further criticized Aristotle's idea whereby every unique soul requires the existence of a unique source, in this case the heart. al-Nafis concluded that "the soul is related primarily neither to the spirit nor to any organ, but rather to the entire matter whose temperament is prepared to receive that soul," and he defined the soul as nothing other than "what a human indicates by saying "I".

## Thomas Aquinas

Following Aristotle and Avicenna, Thomas Aquinas (1225–74) understood the soul to be the first actuality of the living body. Consequent to this, he distinguished three orders of life: plants, which feed and grow; animals, which add sensation to the operations of plants; and humans, which add intellect to the operations of animals.

Concerning the human soul, his epistemological theory required that, since the knower becomes what he knows, the soul is definitely not corporeal—if it is corporeal when it knows what some corporeal thing is, that thing would come to be within it. Therefore, the soul has an operation which does not rely on a body organ, and therefore the soul can exist without a body. Furthermore, since the rational soul of human beings is a subsistent form and not something made of matter and form, it cannot be destroyed in any natural process. The full argument for the immortality of the soul and Aquinas' elaboration of Aristotelian theory is found in Question 75 of the First Part of the Summa Theologica.

## Immanuel Kant

In his discussions of rational psychology, Immanuel Kant (1724–1804) identified the soul as the "I" in the strictest sense, and argued that the existence of inner experience can neither be proved nor disproved. "We cannot prove a priori the immateriality of the soul, but rather only so much: that all properties and actions of the soul cannot be recognized from materiality". It is from the "I", or soul, that Kant proposes transcendental rationalization, but cautions that such rationalization can only determine the limits of knowledge if it is to remain practical.

## Philosophy of Mind

Gilbert Ryle's ghost-in-the-machine argument, which is a rejection of Descartes' mind-body dualism can provide a contemporary understanding of the soul/mind, and the problem concerning its connection to the brain/body.

## James Hillman

Psychologist James Hillman's archetypal psychology is an attempt to restore the concept of the soul, which Hillman viewed as the "self-sustaining and imagining substrate" upon which consciousness rests. Hillman described the soul as that "which makes meaning possible, [deepens] events into experiences, is communicated in love, and has a religious concern", as well as "a special relation with death". Departing from the Cartesian dualism "between outer tangible reality and inner states of mind", Hillman takes the Neoplatonic stance that there is a "third, middle position" in which soul resides. Archetypal psychology acknowledges this third position by attuning to, and often accepting, the archetypes, dreams, myths, and even psychopathologies through which, in Hillman's view, soul expresses itself.

# Religious Views

## Ancient Near East

In the ancient Egyptian religion, an individual was believed to be made up of various elements, some physical and some spiritual.

The souls of Pe and Nekhen towing the royal bargue on a relief of Ramesses II's temple in Abydos.

Similar ideas are found in ancient Assyrian and Babylonian religion. Kuttamuwa, an 8th-century BC royal official from Sam'al, ordered an inscribed stele erected upon his death. The inscription requested that his mourners commemorate his life and his afterlife with feasts "for my soul that is in this stele". It is one of the earliest references to a soul as a separate entity from the body. The 800-pound (360 kg) basalt stele is 3 ft (0.91 m) tall and 2 ft (0.61 m) wide. It was uncovered in the third season of excavations by the Neubauer Expedition of the Oriental Institute in Chicago, Illinois.

## Bahá'í

The Bahá'í Faith affirms that "the soul is a sign of God, a heavenly gem whose reality the most learned of men hath failed to grasp, and whose mystery no mind, however acute, can ever hope to unravel". Bahá'u'lláh stated that the soul not only continues to live after the physical death of the human body, but is, in fact, immortal. Heaven can be seen partly as the soul's state of nearness to God; and hell as a state of remoteness from God. Each state follows as a natural consequence of individual efforts, or the lack thereof, to develop spiritually. Bahá'u'lláh taught that individuals have no existence prior to their life here on earth and the soul's evolution is always towards God and away from the material world.

## Buddhism

Buddhism teaches that all things are in a constant state of flux: all is changing, and no permanent state exists by itself. This applies to human beings as much as to anything else in the cosmos. Thus, a human being has no permanent self. According to this doctrine of *anatta* (Pāli; Sanskrit: *anātman*) – "no-self" or "no soul" – the words "I" or "me" do not refer to any fixed thing. They are simply convenient terms that allow us to refer to an ever-changing entity.

The *anatta* doctrine is not a kind of materialism. Buddhism does not deny the existence of "immaterial" entities, and it (at least traditionally) distinguishes bodily states from mental states. Thus, the conventional translation of *anatta* as "no-soul" can be confusing. If the word "soul" simply refers to an incorporeal component in living things that can continue after death, then Buddhism does not deny the existence of the soul. Instead, Buddhism denies the existence of a permanent entity that remains constant behind the changing corporeal and incorporeal components of a living being. Just as the body changes from moment to moment, so thoughts come and go, and there is no permanent state underlying the mind that experiences these thoughts, as in Cartesianism. Conscious mental states simply arise and perish with no "thinker" behind them. When the body dies, Buddhists believe the incorporeal mental processes continue and are reborn in a new body. Because the mental processes are constantly changing, the being that is reborn is neither entirely different from, nor exactly the same as, the being that died. However, the new being is *continuous* with the being that died – in the same way that the "you" of this moment is continuous with the "you" of a moment before, despite the fact that you are constantly changing.

Buddhist teaching holds that a notion of a permanent, abiding self is a delusion that is one of the causes of human conflict on the emotional, social, and political levels. They add that an understanding of *anatta* provides an accurate description of the human condition, and that this understanding allows us to pacify our mundane desires.

Various schools of Buddhism have differing ideas about what continues after death. The Yogacara school in Mahayana Buddhism said there are Store consciousness which continue to exist after death. In some schools, particularly Tibetan Buddhism, the view is that there are three minds: *very subtle mind*, which does not disintegrate in death; *subtle mind*, which disintegrates in death and which is "dreaming mind" or "unconscious mind"; and *gross mind*, which does not exist when one is *sleeping*. Therefore, *gross mind* is less permanent than subtle mind, which does not exist in death. *Very subtle mind*, however, does continue, and when it "catches on", or coincides with phenomena, again, a new *subtle mind* emerges, with its own personality/assumptions/habits, and *that* entity experiences karma in the current continuum.

Plants were said to be non-sentient, but Buddhist monks are required to not cut or burn trees, because some sentient beings rely on them. Some Mahayana monks said non-sentient beings such as plants and stones have buddha-nature.

Certain modern Buddhists, particularly in Western countries, reject—or at least take an agnostic stance toward—the concept of rebirth or reincarnation, which they view as incompatible with the concept of *anatta*. Stephen Batchelor discusses this issue in his book *Buddhism Without Beliefs*. Others point to research that has been conducted at the University of Virginia as proof that some people are reborn.

## Christianity

Most Christians understand the soul as an ontological reality distinct from, yet integrally connected with, the body. Its characteristics are described in moral, spiritual, and philosophical terms. Richard Swinburne, a Christian philosopher of religion at Oxford University, wrote that "it is a frequent criticism of substance dualism that dualists cannot say what souls are. Souls are immaterial subjects of mental properties. They have sensations and thoughts, desires and beliefs, and perform intentional actions. Souls are essential parts of human beings". According to a common Christian eschatology, when people die, their souls will be judged by God and determined to go to Heaven or to Hell. Though all branches of Christianity – Catholics, Eastern Orthodox, Oriental Orthodox, Evangelical and mainline Protestants teach that Jesus Christ plays a decisive role in the Christian salvation process, the specifics of that role and the part played by individual persons or ecclesiastical rituals and relationships, is a matter of wide diversity in official church teaching, theological speculation and popular practice. Some Christians believe that if one has not repented of one's sins and has not trusted in Jesus Christ as Lord and Savior, he/she will go to Hell and suffer eternal damnation or eternal separation from God. Some hold a belief that babies (including the unborn) and those with cognitive or mental impairments who have died will be received into Heaven on the basis of God's grace through the sacrifice of Jesus.

Soul carried to Heaven by William Bouguereau

Other Christians understand the soul as the life, and believe that the dead are sleeping (Christian conditionalism). This belief is traditionally accompanied by the belief that the unrighteous soul will cease to exist instead of suffering eternally (annihilationism).

Believers will inherit eternal life either in Heaven, or in a Kingdom of God on earth, and enjoy eternal fellowship with God.

There are also beliefs in universal salvation.

*The Damned Soul.* Drawing by Michelangelo Buonarroti c. 1525

## Trichotomy of the Soul

Augustine, one of western Christianity's most influential early Christian thinkers, described the soul as "a special substance, endowed with reason, adapted to rule the body". Some Christians espouse a trichotomic view of humans, which characterizes humans as consisting of a body (*soma*), soul (*psyche*), and spirit (*pneuma*). However, the majority of modern Bible scholars point out how spirit and soul are used interchangeably in many biblical passages, and so hold to dichotomy: the view that each of us is body and soul. Paul said that the "body wars against" the soul, "For the word of God is living and active and sharper than any two-edged sword, and piercing as far as the division of soul and spirit" (Heb 4:12 NASB), and that "I buffet my body", to keep it under control. Trichotomy was changed to dichotomy as tenet of Christian faith at the Council of Constantinople in 869 regarded as the 8th Ecumenical Council by Roman Catholics.

## Origin of the Soul

The 'origin of the soul' has provided a vexing question in Christianity. The major theories put forward include soul creationism, traducianism and pre-existence. According to creationism, each individual soul is created directly by God, either at the moment of conception or some later time. According to traducianism, the soul comes from the parents by natural generation. According to the preexistence theory, the soul exists before the moment of conception. There have been differing thoughts regarding whether human embryos have souls from conception, or there is a point between conception and birth where the fetus acquires a soul, consciousness, and/or personhood. Stances in this question might more or less influence judgements on the morality of abortion.

## Various Denominations

The present Catechism of the Catholic Church defines the soul as "the innermost aspect of humans, that which is of greatest value in them, that by which they are in God's image described as 'soul' signifies the *spiritual principle* in man". All souls living and dead will be judged by Jesus Christ when he comes back to earth. The Catholic Church teaches that the existence of each individual soul is dependent wholly upon God: "The doctrine of the faith affirms that the spiritual and immortal soul is created immediately by God."

Depiction of the soul on a 17th-century tombstone at the cemetery of the Old Dutch Church of Sleepy Hollow

Protestants generally believe in the soul's existence, but fall into two major camps about what this means in terms of an afterlife. Some, following Calvin, believe in the immortality of the soul and conscious existence after death, while others, following Luther, believe in the mortality of the soul and unconscious "sleep" until the resurrection of the dead. Various new religious movements derived from Adventism—including Christadelphians, Seventh-day Adventists and Jehovah's Witnesses—similarly believe that the dead do not possess a soul separate from the body and are unconscious until the resurrection.

The Church of Jesus Christ of Latter-day Saints teaches that the spirit and body together constitute the Soul of Man (Mankind). "The spirit and the body are the soul of man." Latter-day Saints believe that the soul is the union of a pre-existing, God-made spirit and a temporal body, which is formed by physical conception on earth. After death, the spirit continues to live and progress in the Spirit world until the resurrection, when it is reunited with the body that once housed it. This reuniting of body and spirit results in a perfect soul that is immortal and eternal and capable of receiving a fulness of joy. Latter-day Saint cosmology also describes "intelligences" as the essence of consciousness or agency. These are co-eternal with God, and animate the spirits. The union of a newly created spirit body with an eternally-existing intelligence constitutes a "spirit birth" and justifies God's title "Father of our spirits".

## Hinduism

Hindu last rites for departed souls

In Hinduism, the Sanskrit words most closely corresponding to soul are *jiva*, *Ātman* and "purusha", meaning the individual self. The term "soul" is misleading as it implies an object possessed, whereas Self signifies the subject which perceives all objects. This Self (Ātman) is held to be distinct from the various mental faculties such as desires, thinking, understanding, reasoning and self-image (ego), all of which are considered to be part of *prakriti* (nature).

The three major schools of Hindu philosophy agree that the *jiva* (individual self) is related to Brahman (Self/God) but they differ in their explanations of the nature of this relationship. In Advaita Vedanta, the individual self and the Self are deemed to be one and the same. Dvaita rejects this concept of identity and instead identifies the self (jiva) as a separate but similar part of Self, that only becomes one with the Absolute Atman upon Self Realisation (Moksha). Visishtadvaita explains that the nature of the self (jiva) and Brahman (God) is that the jiva is God with attributes; this essentially means that the jiva is both different and the same as God concurrently. Visishtadvaita (Attributive Monism) is thus similar to *Achintya Bhedābeda Tattva* (Inconceivable Difference-Non difference Reality).

The *jiva* becomes involved in the process of becoming and transmigrating through cycles of birth and death because of ignorance of its own true nature. The spiritual path consists of self-realization – a process in which one acquires the knowledge of the self (*brahma-jñanam*) and through this knowledge applied through meditation and realization one then returns to the Source which is Brahman.

The qualities which are common to both Brahman and atmam are being (*sat*), consciousness (*chit*), and bliss/love (*ananda*). Liberation or *moksha* is liberation from all limiting adjuncts (*upadhis*) and the unification with Brahman.

The Mandukya Upanishad verse 7 describes the *atman* in the following way:

"Not inwardly cognitive, not outwardly cognitive, not both-wise cognitive, not a cognition-mass, not cognitive, not non-cognitive, unseen, with which there can be no dealing, ungraspable, having no distinctive mark, non-thinkable, that cannot be designated, the essence of the assurance of which is the state of being one with the Self, the cessation of development, tranquil, benign, without a second (a-dvaita)—[such] they think is the fourth. That is the Self. That should be discerned."

In Bhagavad Gita 2.20 Lord Krishna describes the *atman* in the following way:

na jayate mriyate va kadacin 'nayam bhutva bhavita va na bhuyah 'ajo nityah sasvato yam purano 'na hanyate hanyamane sarire

"For the *atman* there is neither birth nor death at any time. He has not come into being, does not come into being, and will not come into being. He is unborn, eternal, ever – existing and primeval. He is not slain when the body is slain". [Translation by A. C. Bhaktivedanta Swami Prabhupada (Srila Prabhupada)]

Srila Prabhupada, a great Vaishnava saint of the modern time further explains: "The *atman* does not take birth there, and the atman does not die... And because the atman has no birth, he therefore has no past, present or future. He is eternal, ever-existing and primeval – that is, there is no trace in history of his coming into being."

Since the quality of Atma is primarily consciousness, all sentient and insentient beings are pervaded by Atma, including plants, animals, humans and gods. The difference between them is the contracted or expanded state of that consciousness. For example, animals and humans share in common the desire to live, fear of death, desire to procreate and to protect their families and territory and the need for sleep, but animals' consciousness is more contracted and has less possibility to expand than does human consciousness.

When the Atma becomes embodied it is called birth, when the Atma leaves a body it is called death. The Atma transmigrates from one body to another body based on karmic [performed deeds] reactions.

In Hinduism, the Sanskrit word most closely corresponding to soul is Atma, which can mean soul or even God. It is seen as the portion of Brahman within us. Hinduism contains many variant beliefs on the origin, purpose, and fate of the atma. For example, advaita or non-dualistic conception of the atma accords it union with Brahman, the absolute uncreated (roughly, the Godhead), in eventuality or in pre-existing fact. Dvaita or dualistic concepts reject this, instead identifying the atma as a different and incompatible substance.

There are 25 coverings wrapped on our Atma (Reference Taken from *Vaikunta Varnane* written by Sanyasi Vadiraja Swami) 1. Iccha avarka, 2. Linga deha, 3. Avyakta Sharira, 4. Avidya Avarna, 5. Karma avarna, 6. Kama avarna, 7. Jeevacchadaka, 8. Paramacchadaka, 9. Narayana rupa avarna, 10. Vasudeva rupa Avarna, 11. Sankarshana rupa avarna, 12. Pradhyumna Avarka, 13. Anniruddha avarka, 14. Anniruddha Sharira, 15. Vasudeva Kavaca, 16. Narayana Kavaca, 17. Anandamaya kosha, 18. Vignanamaya kosha, 19. Manomaya kosha, 20. Vangmaya kosha, 21. Shrotrumaya kosha, 22. Chakshurmaya kosha, 23. Pranamaya kosha, 24. Annamaya kosha, 25. Gross Body.

According to Brahma Kumaris, soul is an eternal point of light, resides between forehead.

## Islam

The Qur'an mentions two concepts of Ruh (Spirit) and Nafs (Psyche).

Islam teaches that the spirit (Ruh) is immortal and eternal, and that what a person does is recorded and will be judged at the final court of God. They will either go to heaven or hell, depending on whether or not they did well in the test that was given to them by Allah.

> And they ask you, [O Muhammad], about the spirit (Rûh). Say, "The spirit (Rûh) is of the affair of my Lord. And mankind have not been given of knowledge except a little." - Qur'an 17:85

> It is Allah that takes the Rûh at death: and those that die not (He takes their Rûh) during their sleep: then those on whom He has passed the Decree of death He keeps back (their Rûh from returning to their bodies); but the rest He sends (their Rûh back to their bodies) for a term appointed. Verily in this are Signs for those who contemplate. - Qur'an 39:42

## Jainism

In Jainism every living being, from a plant or a bacterium to human, has a soul and the concept forms the very basis of Jainism. The soul *(Jīva)* is basically categorized in two based on its present state.

1. Liberated Souls- These are souls which have attained liberation (Moksha) and never become part of the life cycle again.

2. Non-Liberated Souls - The souls of any living being which are stuck in the life cycle of 4 forms Manushya Gati (Human Being), Tiryanch Gati (Any other living being), Dev Gati (Heaven) and Narak Gati (Hell). Till the time the soul is not liberated from the *Saṃsāra* (cycle of repeated birth and death), it gets attached to different types of above bodies based on the karma (actions) of individual

soul. According to Jainism, there is no beginning and end to the existence of soul. It is eternal in nature and changes its form till it attains liberation.

Irrespective of which state the soul is in, it has got the same attributes and qualities. The difference between the liberated and non-liberated souls is that the qualities and attributes are manifested completely in case of *Siddha* (liberated soul) as they have overcome all the karmic bondages whereas in case of non-liberated souls they are partially exhibited.

Concerning the Jain view of the soul, Virchand Gandhi quoted "the soul lives its own life, not for the purpose of the body, but the body lives for the purpose of the soul. If we believe that the soul is to be controlled by the body then soul misses its power".

## Judaism

The fruit of a righteous man is the tree of life, and the wise man acquires souls.

*Proverbs 11:30*

The Hebrew terms *nephesh* (literally "living being"), *ruach* (literally "wind"), *neshama* (literally "breath"), *chaya* (literally "life") and *yechidah* (literally "singularity") are used to describe the soul or spirit. In Judaism the soul is believed to be given by God to a person as mentioned in Genesis, "And the LORD God formed man [of] the dust of the ground, and breathed into his nostrils the breath of life; and man became a living being." Genesis 2:7. Judaism relates the quality of one's soul to one's performance of the commandments, *mitzvot*, and reaching higher levels of understanding, and thus closeness to God. A person with such closeness is called a *tzadik*. Therefore, Judaism embraces the commemoration of the day of one's death, *nahala/Yahrtzeit* and not the birthday as a festivity of remembrance, for only toward the end of life's struggles, tests and challenges could human souls be judged and credited - *b'ezrat HaShem* ("with God's help") - for righteousness and holiness. Judaism places great importance on the study of the souls.

For I [Hashem] will not contend forever, neither will I be wroth to eternity, when a spirit from before Me humbles itself, and חוּר_ souls [which] I have made.

*Nevi'im, Isaiah 57:16*

Kabbalah and other mystic traditions go into greater detail into the nature of the soul. Kabbalah separates the soul into five elements, corresponding to the five worlds:

1. *Nephesh*, related to natural instinct.
2. *Ruach*, related to emotion and morality.
3. *Neshamah*, related to intellect and the awareness of God.

4. *Chayah*, considered a part of God, as it were.
5. *Yechidah*, also termed the *pintele Yid* (the "essential [inner] Jew"). This aspect is essentially one with God.

Kabbalah also proposed a concept of reincarnation, the *gilgul*.

## Scientology

The Scientology view is that a person IS a soul. They do not 'have' a soul. A person is immortal, and may be reincarnated if they wish. The Scientology term for the soul is "thetan", derived from the Greek word "theta", symbolizing thought. Scientology counselling (called 'auditing') addresses the soul to improve abilities, both worldly and spiritual.

## Shamanism

According to Nadya Yuguseva, a shaman from the Altai, "A woman has 40 souls; men have just one"

## Sikhism

Sikhism considers Soul (atma) to be part of God (Waheguru). Various hymns are cited from the holy book "Sri Guru Granth Sahib" (SGGS) that suggests this belief. "God is in the Soul and the Soul is in the God." The same concept is repeated at various pages of the SGGS. For example: "The soul is divine; divine is the soul. Worship Him with love." and "The soul is the Lord, and the Lord is the soul; contemplating the Shabad, the Lord is found." The "Atma" or "Soul" according to Sikhism is an entity or "spiritual spark" or "light" in our body because of which the body can sustain life. On the departure of this entity from the body, the body becomes lifeless – No amount of manipulations to the body can make the person make any physical actions. The soul is the 'driver' in the body. It is the 'roohu' or spirit or atma, the presence of which makes the physical body alive. Many religious and philosophical traditions, support the view that the soul is the ethereal substance – a spirit; a non material spark – particular to a unique living being. Such traditions often consider the soul both immortal and innately aware of its immortal nature, as well as the true basis for sentience in each living being. The concept of the soul has strong links with notions of an afterlife, but opinions may vary wildly even within a given religion as to what happens to the soul after death. Many within these religions and philosophies see the soul as immaterial, while others consider it possibly material.

## Taoism

According to Chinese traditions, every person has two types of soul called hun and po, which are respectively yang and yin. Taoism believes in ten souls, *sanhunqipo* "three *hun* and seven *po*". The pò is linked to the dead body and the grave, whereas the hún is linked to the ancestral tablet. A living being that loses any of them is said to have mental illness or unconsciousness, while a dead soul may reincarnate to a disability, lower desire realms or may even be unable to reincarnate.

## Zoroastrianism

## Other Religious Beliefs and Views

Charon who guides dead souls to the Underworld. 4th century BC

In theological reference to the soul, the terms "life" and "death" are viewed as emphatically more definitive than the common concepts of "biological life" and "biological death". Because the soul is said to be transcendent of the *material existence,* and is said to have (potentially) eternal life, the death of the soul is likewise said to be an *eternal death.* Thus, in the concept of divine judgment, God is commonly said to have options with regard to the dispensation of souls, ranging from Heaven (i.e., angels) to hell (i.e., demons), with various concepts in between. Typically both Heaven and hell are said to be eternal, or at least far beyond a typical human concept of lifespan and time.

## Spirituality, New Age and New Religions

## Brahma Kumaris

In Brahma Kumaris, human souls are believed to be incorporeal and eternal. God is considered to be the Supreme Soul, with maximum degrees of spiritual qualities, such as peace, love and purity.

## Theosophy

In Helena Blavatsky's Theosophy, the soul is the field of our psychological activity (thinking, emotions, memory, desires, will, and so on) as well as of the so-called paranormal or psychic phenomena (extrasensory perception, out-of-body experiences, etc.). However, the soul is not the highest, but a middle dimension of human beings. Higher than the soul is the spirit, which is considered to be the real self; the source of everything we call "good"—happiness, wisdom, love, compassion, harmony, peace, etc. While the spirit is eternal and incorruptible, the soul is not. The soul acts as a link between the material body and the spiritual self, and therefore shares some characteristics of both. The soul can be attracted either towards the spiritual or towards the material realm, being thus the "battlefield" of good and evil. It is only when the soul is attracted towards the spiritual and merges with the Self that it becomes eternal and divine.

## Anthroposophy

Rudolf Steiner differentiated three stages of soul development, which interpenetrate one another in consciousness:

- The "sentient soul", centering on sensations, drives, and passions, with strong conative (will) and emotional components;
- The "intellectual" or "mind soul", internalizing and reflecting on outer experience, with strong affective (feeling) and cognitive (thinking) components; and
- The "consciousness soul", in search of universal, objective truths.

## Miscellaneous

In Surat Shabda Yoga, the soul is considered to be an exact replica and spark of the Divine. The purpose of Surat Shabd Yoga is to realize one's True Self as soul (Self-Realisation), True Essence (Spirit-Realisation) and True Divinity (God-Realisation) while living in the physical body.

Similarly, the spiritual teacher Meher Baba held that "Atma, or the soul, is in reality identical with Paramatma the Oversoul — which is one, infinite, and eternal...[and] [t]he sole purpose of creation is for the soul to enjoy the infinite state of the Oversoul consciously."

Eckankar, founded by Paul Twitchell in 1965, defines Soul as the true self; the inner, most sacred part of each person.

## Science

The findings of science may be relevant to one's understanding of the soul depending on one's belief regarding the relationship between the soul and the mind. Another may be one's belief regarding the relationship between the soul and the body. Science has never proven the existence of a soul.

## Neuroscience and the Soul

Neuroscience as an interdisciplinary field, and its branch of cognitive neuroscience particularly, operates under the ontological assumption of physicalism. In other words, it assumes—in order to perform its science—that only the fundamental phenomena studied by physics exist. Thus, neuroscience seeks to understand mental phenomena within the framework according to which human thought and behavior are caused solely by physical processes taking place inside the brain, and it operates by the way of reduction by seeking an explanation for the mind in terms of brain activity.

To study the mind in terms of the brain several methods of functional neuroimaging are used to study the neuroanatomical correlates of various cognitive processes that constitute the mind. The evidence from brain imaging indicates that all processes of the mind have physical correlates in brain function. However, such correlational studies cannot determine whether neural activity plays a causal role in the occurrence of these cognitive processes (correlation does not imply causation) and they cannot determine if the neural activity is either necessary or sufficient for such processes to occur. Identification of causation, and of necessary and sufficient conditions requires explicit experimental manipulation of that activity. If manipulation of brain activity changes consciousness, then a causal role for that brain activity can be inferred. Two of the most common types of manipulation experiments are loss-of-function and gain-of-function experiments. In a loss-of-function (also called "necessity") experiment, a part of the nervous system is diminished or removed in an attempt to determine if it is necessary for a certain process to occur, and in a gain-of-function (also called "sufficiency") experiment, an aspect of the nervous system is increased relative to normal. Manipulations of brain activity can be performed with direct electrical brain stimulation, magnetic brain stimulation using transcranial magnetic stimulation, psychopharmacological manipulation, optogenetic manipulation and by studying the symptoms of brain damage (case studies) and lesions. In addition, neuroscientists are also investigating how the mind develops with the development of the brain.

## Physics and the Soul

Physicist Sean M. Carroll has written that the idea of a soul is in opposition to quantum field theory (QFT). He writes that for a soul to exist: "Not only is new physics required, but dramatically new physics. Within QFT, there can't be a new collection of 'spirit particles' and 'spirit forces' that interact with our regular atoms, because we would have detected them in existing experiments."

In contrast, Princeton Professor Hans Halvorson has highlighted a conceptual difficulty in making quantum mechanics logically consistent. His 2016 debate with Sean Carroll at Caltech on this subject is available from several sources on the web. Halvorson points out that some interpretations of the measurement problem in quantum mechanics seem to require an observation by a mind or conscious agent to collapse the wave function to obtain a determinate result. Since quantum mechanics posits a superposition as the normal state of matter, a material brain should also be in an indeterminate state much like Schrödinger's Cat. How then can it produce a determinate result? Halvorson argues that this logically necessitates something very like an immaterial soul that can make the observation and collapse the wave function. This general problem was first pointed out by physicist Eugene Wigner, who thought wave function collapse occurred due to the activities of mind. Theoretical physicist Roger Penrose and others have developed a similar outlook.

Quantum indeterminism has been invoked by some theorists as a solution to the problem of how a soul might interact with the brain but neuroscientist Peter Clarke found errors with this viewpoint, noting there is no evidence that such processes play a role in brain function; and concluded that a Cartesian soul has no basis from quantum physics.

### Biology and the Soul

Biologist Cyrille Barrette (fr) has written that "the soul is a word to designate an idea we invented to represent the sensation of being inhabited by an existence, by a conscience". Barrette explains, using simple examples in a short self-published article, that the soul is a property emerging from the complex organisation of matter in the brain.

### Parapsychology

Some parapsychologists have attempted to establish, by scientific experiment, whether a soul separate from the brain exists, as is more commonly defined in religion rather than as a synonym of psyche or mind. Milbourne Christopher (1979) and Mary Roach (2010) have argued that none of the attempts by parapsychologists have yet succeeded.

### Weight of the Soul

In 1901 Duncan MacDougall made weight measurements of patients as they died. He claimed that there was weight loss of varying amounts at the time of death, he concluded the soul weighed 21 grams. The physicist Robert L. Park has written that MacDougall's experiments "are not regarded today as having any scientific merit" and the psychologist Bruce Hood wrote that "because the weight loss was not reliable or replicable, his findings were unscientific."

## Concept of Plato

Soul has the power of performing cognitive activity. As knowledge presupposes a knower, the soul plays the role of a knower. It is the subject of knowledge. The process of knowledge acquisition is the matter of our study here to explore the kind of dualism

Plato presupposes in his epistemology. As we have mentioned earlier that this engagement is an epistemic engagement which shows the relationship between the knower and the known. The known here signifies *the object* of knowledge. In Plato, this subject (Knower) and object (the Known) dichotomy is not only epistemological but also ontological. These two forms of dichotomy are in fact related. The epistemological dichotomy refers to the division that Plato maintains between the *forms/ ideas* and the *particulars*. The forms represent the essence of things. For example, *horseness* is the *form* of all horses. Horseness *participates* in all existing particular horses in the world, irrespective of their physical appearances and differences in terms of their size, colour, gender, and ability. The particulars are thus the copies of the *forms* which according to Plato participate in all the particulars – *things in the world*. Forms are universal and don't change their ontological status. Forms are the basic source of knowledge. The knowledge about the particulars is possible because of the existence of forms.

The forms are independent of particulars and *akin* to the soul through which the existence of the particulars is being conceived. For instance, to know that 'this is a horse' is to have *idea* (*knowledge of form*) about the horse. The soul is conscious and hence is able to conceive the *ideas*. Ideas/ forms, according to Plato exist in an independent world. Their participation in the particulars does not make them dependent on the particulars. Rather, the particulars exist in the world. The knowledge of the particulars is not only dependent on the existence of ideas/ forms but also derived from sense experiences. The sense experiential knowledge varies from person to person. So, the knowledge of particulars can be *subjective*, whereas the knowledge of forms is *objective*. Thus, briefly the world of forms/ ideas is different from the world of particulars. The forms and the particulars belong to two different worlds. Hence, they have two different ontological statuses.

The soul is the cause of the *motion* (bodily movements) or the action (voluntary/ intentional action). Plato compares the nature of the soul with the nature of the universe. Soul comprises of two aspects: the subject of knowledge or the source of cognitive activity, and the principle of movement. In Greek philosophy soul is defined as the *vital force*. It is the center of intelligent personality. The soul is autonomous and is independent of the world. The soul is the source of all kind of cognitive activities. The nature physical activities are different from the cognitive activities. Nevertheless, the soul is subject of performing or initiating actions. So far as the cognitive activities are concerned, it experiences thought, emotions and so on. The aspect of the soul departs at the time of the death. Plato here mentions of the *motion* and the *cognition* as two aspects. Modes of action mean the recognition of cognitive action or thought as one of the important activities of the soul. As the self performs spontaneous action, it is considered to be a self moving reality. This motion is rational. Nevertheless, soul is an intermediary between the two realms of the existence. Theory of being (the soul) was an unexamined presumption. Soul is thus a metaphysical entity that intermediates between two realms of the existence; the eternal Ideas/ *forms* and the mundane particulars. Thus the soul is not absolutely merged with the world.

In this way, we trace from Plato's ontological dualism from the discourse of epistemology, particularly when he delve into the study of the nature of Forms/ Ideas with reference to Soul. Knowledge, according to Plato, is an unchanging reality. It is universally true which is intuitively conceived by the soul. And, knowledge of the changing world, i.e., ceaseless flow of the particulars in the world, is known through the senses organs. There are two kinds of knowledge – knowledge through sense experience and knowledge through intuition. The *knowledge of forms* is *intuitive* knowledge. In this connection, Plato makes a distinction between *doxa* (opinion) and the knowledge. "Opinions are based on sense perception. The changing objects in the perception are relatively unreal".

## Ideas are Knowable Unchanging Objects of the Knowledge

Knowledge of particulars may vary from context to context; hence there is a possibility that such impression can create *opinion* means mere beliefs that are relatively true. In this context, Plato maintains that 'knowledge is justified true belief.' Justified belief involves certainty whereas an unjustified belief is uncertain and relative.[1] Opinions are often formed without any proper justification. Hence, opinions are not regarded as valid knowledge. From the above discussion we can also conclude that the soul is a third kind of entity. It is the presumption of knowledge. Knowledge of the soul is a transcendental knowledge. Soul is the central to Plato's epistemology as well as metaphysics. The true nature of forms is grasped by gradually transcending the experiences of the empirical world. The empirical world, for Plato, is the copy of the world of Forms. Forms themselves are real and universal. They participate in the particulars. This, however, does not eliminate the dualism in Plato's philosophy. Rather, this dualism of *Form* and *Content* reappears again and again because of the abstract opposition between these two realms.

According to Plato, the soul is 'akin' to ideas or forms. This relationship is defined as an affinity. Affinity is a condition by the virtue of which the soul contemplates or apprehends true being. Moreover Plato puts emphasis on the ethical as well as the ontological considerations which tend to deny that there is an essential relationship between the soul and the objects of the sense experience.

## Ethics and Ontology

So far as ethics and ontology are concerned, Plato emphasizes that truth and good are one. Knowledge is identical with the virtue. In the hierarchy of virtues, Plato places goodness at the highest level. Goodness is the supreme virtue. Every individual must live a virtuous life – the life must be expressed in performing with good actions. Along with the form of Good, Plato did also talk about Justice and Beauty which are in harmony with each other so far as living the virtuous life is concerned. Hence, it is essential to have the knowledge of the form. Referring to the autonomy of Forms, Mackie writes, "In Plato's theory of Forms, and in the particular the Form of the Good, are eternal, extra mental, realities. They are a very central structural element in the fabric of the world. But it is held also that just knowing them or 'seeing' them will not merely tell men what to do but will ensure that they do it, overruling any contrary inclinations. The philosopher-kings in the *Republic* can, Plato thinks, be trusted with unchecked power because their education will have given them knowledge of the Forms. Being acquainted with Forms of the Good, and justice and beauty and the rest they will, by this knowledge alone, without any further motivation, be impelled to pursue and promote these ideals." Plato's emphasis on the epistemic articulation of the ethical highlights the how moral actions are ought to be performed and what short of life the human beings are ought to live. The cognitive activity of the soul is to contemplate on the ideas.

Soul works with the bodily facilitation as well. Soul is neither a divisible substance nor it is concrete. It has got a unique cognitive status in Plato's epistemology. The nature of the cognitive activity is epistemological but not psychological. The epistemological activities are concerned with truth, whereas the psychological is not concerned with truth.Moreover, the soul's engagement with the external reality is something significant, but so far as the epistemology is concerned, the soul deals with the understanding of one's own reality. That is, the power of conceiving its own existence is indeed a unique cognitive act performed by the soul. The knowledge of the soul is not derived through sense organs. For example; to be good and have the knowledge of good are two different things. One can just perform good action accidently without knowing the significance. On the other hand, the wise person not only possesses the knowledge of good, but also cultivates the virtue of goodness. The seeker of knowledge is a lover of wisdom. Wisdom is the state of the soul. It is pure, eternal, immortal etc.

## Does Soul Perform Action with an Ethical Reason?

The ethical aspect of the ontology talks about living a virtuous life. For Plato, the human beings are ought to live a virtuous life. Virtue is knowledge. Soul as the originator of voluntary action performs action rationally.

The virtues are to rationally construed as the soul has three different parts namely, reason, spirit and appetite. The reason is considered supreme as it has the power to evaluate and reflect on things. Appetite and spirit are *akin* to the body rather than to the soul. The body is a hindrance to the soul, in the sense that it restricts its function to the sense activity and thereby prevents the soul from reaching pure truth.

Goodness is considered supreme Idea. It is represented as an organizing principle in the universe and knowledge. The reason to have a good life and form the habit of living a good life is a virtue. Rational and scientific thinking (knowledge) is stepping stone towards the knowledge of truth.

Immortality of soul in Plato corresponds to divinity and eternity in its knowledge of Ideas. Immortality in the sense of continued existence conceived by Plato as 'personal being' – is the principle of personality. Souls differ according to their respective degree of development. Roberts also mentions that "the other kind of immortality – that to which the soul approximates in virtue of its kinship with ideas – is qualitative rather than durational."

Plato's notion of Ideas is to be understood as eternal, timeless and self explanatory. And, the basic particulars in the world is not entirely or absolutely one with reason.

Roberts writes "He whose heart has been set on the love of learning and on true wisdom, and has chiefly exercised this part on himself, this man must without fail have thoughts that are immortal and divine, if he laid hold upon truth; and so far as it lies in human nature to possess immortality he lacks nothing thereof."

## In the Republic

In Book IV of the *Republic* Plato and his interlocutors are attempting to answer whether the soul is one or made of parts. Plato states that, "It is clear that the same thing will never do or undergo opposite things in the same part of it and towards the same thing at the same time; so if we find this happening, we shall know it was not one thing but more than one." (This is an example of Plato's Principle of Non-Contradiction.) For instance, it seems that, given each person has only one soul, it should be impossible for a person to simultaneously desire something yet also at that very moment be averse to the same thing, as when one is tempted to commit a crime but also averse to it. Both Plato and Glaucon agree that it should not be possible for the soul to at the same time both be in one state and its opposite. From this it follows that there must be at least two aspects to soul.

## Logical

The logical or *logistikon* (from *logos*) is the thinking part of the soul which loves the truth and seeks to learn it. Plato originally identifies the soul dominated by this part with the Athenian temperament. The *logistikon* discerns what is the real and not merely apparent, judges what is true and what is false and wisely makes just decisions in accordance with its love for goodness.

Plato makes the point that the *logistikon* would be the smallest part of the soul (as the rulers would be the smallest population within the Republic), but that, nevertheless, a soul can be declared just only if all three parts agree that the *logistikon* should rule.

## Spirited

According to Plato, the spirited or *thymoeides* (from *thymos*) is the part of the soul by which we are angry or get into a temper. He also calls this part 'high spirit' and initially identifies the soul dominated by this part with the Thracians, Scythians and the people of 'northern regions.' In the just soul, the spirited aligns with the *logistikon* and resists the desires of the appetitive, becoming manifested as 'indignation' and in general the courage to be good. In the unjust soul, the spirited ignores the *logistikon* and aligns with the desires of the appetitive, manifesting as the demand for the pleasures of the body.

## Appetitive

The appetitive or *epithymetikon* (from *epithymia*) is the part of the soul by which we experience carnal erotic love, hunger, thirst and in general the desires opposed to the *logistikon*. (The appetitive is in fact labelled as being 'a-logical'.)

Plato also identifies this part of the soul with the pleasure involved in human reproduction. He further relates this part to the love of money-making, which he mentions as being the particular mark of the Phoenicians and Egyptians.

## Concept of Aristotle

Soul is the *principle of life*. The concept of life is conceptualized looking at different functions of the living organisms. For Aristotle, soul is the *essence* of life and is *coextensive with life*. It signifies the activities of living beings correspond to life of plants, insects, beasts and human beings. All these organisms embody soul. That is, soul bears a set of capacities – power through which the organic and the inorganic activities are performed. In the case of plants, the capacity could be understood in terms of the power of nutrition. The plants have soul but not sensation. In the case of insects and animals we find not only nutritional abilities but also the sense-perception. Whereas, the capacity of thinking, willing, experiencing, performing voluntary action, and imagination are the additional capacities of the soul of human beings along with the power of nutrition and sense perception. The soul of the human beings thus has the capacity to grow, perceive and think. In this connection, one can find how there are different ways in which the life has been actualized.

'Aristotle emphasizes that the actualization of life to be understood in terms of its continuity, rather than differences, between processes in plants than the processes in human beings.' Life itself is an actualization. It is the internal cause of change. Actualization itself is a disposition. In other words, the actualization of life in different forms presupposes that life exists potentially in the very nature of things. Thus, Aristotle draws a distinction between potency and actualization. The actualized resides in the state of potential. Life represents movement from the potential to the actual. For Aristotle, there is a causal process involved in showing the transition from the potential existence of life to its actualization. This causal process is explained with reference to four kinds of causation. They are: the material cause, formal cause, efficient cause and the final cause.

The Material cause refers to matter in its pure state of existence, e.g. wood, clay, etc. Matter, for Aristotle, is necessarily or in itself not something (this). It is basically conceived as stuff, out of which a thing is produced. When something is produced then it gets an individual identity – this is a pot. The shape or the form gives identity to an object. Aristotle denies the division between form and content, which is advocated by Plato. For Plato, forms are independent of particulars (a particular object). Nevertheless the forms only participate in the particulars. They have two different kinds of ontological status, as they exist in two separate realms. However, Aristotle argues that life as a substance is a combination of both matter and form. The matter is potentiality and form is actualization. The conceptualization of the potential structure of the matter, Aristotle introduces the notion of formal cause. That is, the basic structure (pattern)/ form of the matter. Every material body or a particular assumes a structure when it is being made. So the process of making an object or to show that how an object is being made, we need to understand the basic form of the matter. For instance, to make a pot, the potter needs to conceptualize the very form of clay (material cause) out of which a pot can be designed or made. The existing things in the nature have a formal structure. Unlike Plato the primacy to the notion of form is not overemphasized, rather the form is found to be inseparably connected to the material body. However, the change in the material stuff or the body is initiated by an efficient cause. The efficient cause is the source of initiating change. The efficient cause, in the above example of making the pot, is the potter. The potter brings about the thing. According to Aristotle, the role of an efficient cause is significant as it starts the process of bringing change in the very domain of the material body – the stuff. The efficient cause thus intervenes while being engaged in the process of making something. The intervention of the agent persists for a period and then the agent withdraws himself/ herself from the causal process. That is to say the causal process continues to be there till the product is finally being made. Highlighting this aspect of the process of production, Aristotle brings in the notion of final cause which signifies the very purpose or the end of something. In other words, every object is being made with a purpose. The purpose of making a pot lies in the function of the pot.

The concept of *actuality* is present in two levels. For instance, we have discussed earlier that human beings are *knowledge seeking* beings. A human being is therefore a *potential knower*. Human beings possess the ability to learn and express knowledge in words. Expression of knowledge is different from having the potentiality to acquire knowledge. This could be also illustrated with reference to the notion of language acquisition. The ability to learn language is a first level actuality, whereas ability to correct a grammatically error expressions is a second level actuality. At the first level the thoughts are just expressed, and in the second level, the person is *conscious of* the fact that how to speak correctly or to correct a sentence.

However, we need to have some clarity on the nature of *natural bodies* and the *conscious mind*. For Aristotle, every natural body which shares *life* is a *substance*. The natural substance is of two types: Nature in Physics which includes the things that exist in nature, e.g. animals, plants, and simple elements. And, things that exist by *other causes* include artefact, such as, clock, a bed, *etc.* – that exists by art. The human beings are *living natural bodies*. The body and the soul/ *the conscious mind* stand to each other as matter to form. 'The soul is not an attribute of the body; rather the soul is substance in the sense of form of a natural body which potentially partakes of life.' The soul is the *first level of actualization* of the natural body. The organic body only uses its own parts as means of nutrition – 'a plant uses its roots like a mouth to take food.' In the *second level actualization* if the essence or the substance is separated then the object loses its identity. In the living human body the soul is the substance or essence without which the body loses its identity. Examples: the essence of an *axe* lies in the *function of cutting*. Or, the essence of the eye(s) represented in the form of *sight*. Here, *sight* and the *function of cutting* are considered as *substance – the soul* of things, without which the existence of the object is *meaningless*.

According to Aristotle, the *ensouled* things are distinguished by living, and the life is manifested in many forms.

1. Intellect.
2. Perception.
3. Locomotion and rest.
4. Nutritive movement, growth and decay. (plants)

Animals lack intellect whereas the plants have only the capacity of nutritive movement. Distinct species have distinct powers. Miller says, "A complete account of soul must include an account of each powers and in addition explain their interlocking relationships." Aristotle complains that these philosophers (Plato and Pythagoreans) contended that soul could be fitted into a body without explaining the cause or the condition of the body into which it is *allegedly* entered. Each body has a peculiar form. "... soul is the actuality of the body...the actualization of each thing naturally comes to be in that which has the potentiality (for it) and in the appropriate matter." However, for Aristotle, the soul is basically material/ biological. "The body stands to soul as the matter stands to form." The *leading way* is the process of *actualization*, i.e., the process in which *x* becomes y. So, 'the soul is the actualization of the body as the sailor is the actualization of the ship.'

So far as the intellectual ability of human mind is concerned, Aristotle maintains that consciousness is *intrinsic* to the mental life and has *higher order function*. The soul is consciousness. The intrinsic feature of consciousness exhibits the intentional attitude of the mind. Being intentional shows that human beings are not just conscious like other animals are aware of things in the world. Aristotle holds that animals do have the capacity of awareness, i.e., being aware of something, but they 'are not aware of what they are about.' In other words, human beings are aware of their own mental states; aware of the content of the thoughts. This awareness is self-reflexive and immediate. The reflexive form of awareness shows that we are *conscious of* the very *content* of thought. Hence this form of consciousness is not merely intentional, rather being is conscious of its intentional engagement. For instance, *x* is *seeing* the cloud. The act of seeing is an intentional act, could be performed by any conscious beings. But so far as the human perceptual abilities are concerned, the being perceives the fact that it is seeing/ perceiving things.

That is, to be aware of one's own act of perceiving. "Aristotle always asserts that we perceive that we see, never that we see that we see." Seeing is a sensory activities, could be causally explainable, whereas perception is complex intellectual act. In the case of perceptual awareness, the person knows what is taking place. In other words, perceptual awareness is attached to experience which is shown in the case of change that one undergoes while perceiving something. This awareness is *immediate* and *transparent;* higher order function of consciousness. The notion of transparency shows the evidence of direct knowledge. For instance, when 'I perceive could,' the perceptible object is *same as* the activity of perceptible. From this Caston infers 'Aristotle's commitment to some form of direct realism.' Further this notion of transparency also provides clue to understand *reflexive awareness.*

Caston writes, "...reflexive awareness is both intrinsic and immediate, occurring without the intervention of further acts, causal relations, representation or inferences. And, yet there remains a kind of indirectness about it. We are not aware of the act itself in the same way that we are aware of its primary object. Thus, while we can be said to perceive that we see, it will not be exactly like perceiving an object." The impression of an indirect relationship is shown to highlight that conscious mind has two levels of functions. At the primary level it is directly attached to all kinds of functions of external senses, and at the secondary level the mind reflects upon the primary ones. Thus it appears to have some kind of indirect relations, but in reality it is immediate. The notion of immediate awareness is also explainable with reference to the notion of *introspection* in the Cartesian tradition. Introspection is the *inner sense* of the mind; through which one synthesize the various forms of activities. Caston does illustrate on this point that there is an inner sense distinct from the outer function of different activities of the senses. Thus the Aristotle's notion of mind has a sense of *inner* which could be expressed in the form of *reflexive consciousness.*

## Concept of Descartes

Descartes formulated the mechanistic explanation of the phenomena of nature by introducing the *hypothetical* model. For him, hypothesis needs to be formulated to explain the imperceptible causal properties of the perceptible body. The causal properties of the matter include both the material and the sufficient causes. And the combination of the imperceptible particles together would be able to constitute the perceptible body. The imperceptible particles are the micro states of matter and the perceptible particles are the macro states of matter, which could be explained through the basic properties called shape, size, and motion. And the formulation of mechanistic hypothesis for the explanation of natural phenomena is to bridge the gap between our sensory experiences of matter and their causal capacity. As Clarke mentions, "The distinction between our subjective experiences or sensations and their objective causes, between primary and secondary qualities, opens up an epistemic gap that can only be bridged by some other strategy apart from the assumption of resemblance. This strategy is hypothesis, or guesswork. Our guesswork may turn out to be secure and there may eventually be many reasons for thinking that they are as certain as one can hope for ...." Secondary qualities are mostly subjective in their appearances whereas primary qualities are objective. Here the primary qualities of the matter are the basic properties of it, such as its shape, size and motion through which matter can be explained.

Now one may ask, how do we have the ideas of these imperceptible properties of matter? Descartes elaborates it in his famous example of *wax*. Wax in its original state (fresh from the hive) possesses qualities such as sweetness, fragrance of honey, colour, size, etc. If the same piece of wax is heated, then its size expands, its sweetness disappears, and the colour changes. That is, a certain transformation takes place between the time period before heating and after heating. However, we still have the common notion of wax. And during the course of transformation, we know that heat operates in the composite structure of the wax. We don't perceive the movement of the minute parts of the wax during the period of transformation. Still the conception of wax remains because some essential qualities persist and they help us in identifying the wax as wax after its transformation. For Descartes, this notion of wax could be comprehended neither by perception nor by imagination that physics deals with. Rather, the comprehension of this idea is possible by *understanding*. Understanding deals with "pure intellect". As Wood describes, "After the exclusion of sensation and imagination there remains one possibility, namely, that understanding, exercising its powers of conception and judgment for the synthesizing and unifying function which constitutes the self identity of the piece of wax." Understanding is a special mode for formulating such ideas as ascribed to the function of the intellect. The intellect has a unique way of judging the observable facts. It is through its intellect alone that material bodies are comprehended.

Descartes' philosophy of mind emerges from his reflections on nature as a mechanistic system. He locates the beginning of this philosophy in our subjective experiences. As Clarke cites, "where the metaphysics is concerned, we begin with such everyday experiences as the experience of thinking, feeling, moving, etc. Among these experiences, Descartes favors the most accessible and widely available experiences because he hopes thereby to find indubitable foundation."[3] Among these experiences thinking is considered as the most indubitable and simply accessible feature of our knowledge. Thinking or *cogito* is easily comprehensible, because it is *directly accessible* to me – the self. Descartes does not conceptualize the notion of *self* as representation to itself, rather is a *self-evident* truth. It is perceived by *inner sense* of the mind. The perception of the mind is also known as *introspection* which provides certainty to the existence of the self. Thus the knowledge of the existence of the self is an *indubitable* truth. The self evident nature of the experience of this indubitable truth about the existence of the self puts an end to Descartes search for the foundation of knowledge. This is indubitable because *doubting* presupposes *thinking*. From doubting we cannot obtain certainty. And, Descartes used *doubting* as a kind of method to establish truth.

Thinking is a conscious activity. To say that *I think* is to claim that I am conscious of this very activity called thinking. Thus thinking is a *self-conscious* activity. It is an immediate knowledge to me that *I am thinking*. It is different from bodily activities. The bodily activities are performed with the self being conscious of them. Consciousness includes all mental activities, like thoughts, sensations, imaginations, and experiences. Hence, the mind exists as an independent *substance – res cogitans*. "The certainty of my own existence obtains only when I am thinking, when I am conscious, I am, I exist, that is certain." (Copleston, 1994: 91) The Cartesian metaphysics is based on the principle of *cogito* (*I think*) which is most indubitable and certain. As Descartes claims, "I am nothing so long as I think that I am something. So after considering everything very thoroughly, I must finally conclude that proposition, I am, I exist, is necessarily true whenever it is put forward by me and conceived in my mind."Thus thinking or *cogito* lays down the foundation for any sort of knowledge claims, including the mechanistic explanation of the universe.

Consciousness is the basic metaphysical reality in Descartes' philosophy. For him, the very nature of understanding consciousness is different from the understanding of the nature of the body. The body or matter is usually defined as something extended in space that can be perceived, and also has movement in connection with other bodies. It is different from the mind which is neither perceived no is extended in space. Descartes says, "For according to my judgement, power of self-movement, like the power of sensation or of thought was quite foreign to the nature of the body; indeed it was a source of wonder to me that certain bodies were found to contain faculties of this nature." The self-movement describes the independent function of consciousness. It is not dependent upon the other mechanical functions of the body. Thereby consciousness is characterized as a unique feature of human organism. The human organism is not just sensitive to the external world nature but also intelligent enough to think and describe it.

Descartes' metaphysics of mind goes beyond the boundaries of physics and physiology. It situates the mind beyond the finites of mechanism. The physiological structures of a living body, especially of human beings are organized in a certain way so that they bear certain mechanical functions. These functions include digestion, blood circulation and palpitation of the heart, various movements of limbs, etc. Beyond the functions the conscious behaviour of the body is described by the presence of the soul in the living organism. The soul is placed in the *small gland* in the middle of the brain. The brain is taken as the vital part of the body connected with all other parts of the body through the nerve fibers to carry out the sensations. It is the pineal gland that converts the blood into animal spirits. The sprits are kept in the cavities and further they enter into the muscles through the nerve fibers. The flow of spirits varies depending upon the impressions that are carried to the soul. Descartes also says, “And conversely, the mechanism of the body is so constructed that simply by this gland’s being moved in any way by the soul or by any other cause, it drives the surrounding spirits towards the pores of the brain, which direct them through the nerves of the muscles; and in this way the gland makes the spirit move the limbs.”

This mechanical composition of the body no doubt has a certain biological capacity to generate the movement. All these movements of the body are not derived from the soul and do not have any necessary relationship with consciousness. The heat generated by the natural composition of the body, for instance, is unique to the function of the body. Descartes says, “it has been believed, without justification, that our natural heat and all the movements of our bodies depend on the soul; whereas we ought to hold, on the contrary, that the soul takes its leave when we die only because the heat ceases and the organs which bring about bodily movement decay.” The decomposition of the physiological feature of the body is a natural phenomenon in biology. The body is unable to generate heat or other physiological functions like digestion, palpitation, etc. because it loses certain biological capacities. Now the question arises regarding the interaction of the consciousness with the body or of the soul with the body. Soul has two parts; the rational and the sensitive. The rational part of the soul includes thought processing. It is the higher level having other features like, willing, imagining, and having sensory perception. The sensitive level is only related to the bodily function. It is a lower level function because there is no willingness or volition present in this level of consciousness. The body being sensitive to nature is related to this lower level function. Thus, it is the soul or consciousness which acts on the body.

However, the existence of a person is defined by his being conscious. Consciousness signifies the higher order function of thinking, not just being sensitive. The higher order function of consciousness designates self-movement of the soul. This higher order function is not dependent on the function of the bodily movement whereas “thinking is inseparable from me.” Descartes points out, “I am, then, in a strict sense only a thing that thinks, that is, I am a mind or intelligence, or intellect, or reason¾words whose meaning I have been ignorant of until now.” Thinking entails the *real* existence.

Thus, in Descartes, consciousness or thinking is not associated with the physical or the biological function of the body. Therefore it cannot be reduced to the physical properties of the body. The biological function of the body remains different from the function of consciousness. On the other hand, the outer behaviour or movement of the body is manifested due to the active participation of the soul. The active participation reveals the basic features of the soul. Moreover the interaction remains a mystery. The mystery of the body is being conscious or the body as having a conscious mind is not explicable by the mechanistic model. Consciousness is objectively ‘real’. It is autonomous. This is asserted from the first person point of view. I am only aware of my being conscious. I can only realize my acting consciously or thinking consciously. The moment I cease to think, I am unable to comprehend my existence any more. Hence consciousness plays a necessary role for my understanding the world and representing it. Consciousness has to be understood as an irreducible and independent principle of knowledge. It is because neither can it be reduced to a bodily function nor can it be explained sufficiently by bodily function. Consciousness, thus, is metaphysically distinct from the material body. This is the source of the Cartesian mind-body dualism.

## Dualism

The Cartesian theorization of the mental has been immensely criticized. As it fails to show the connection between the mind and the body, the dualism remains open to further debates and criticisms. To talk about this causal relation between the mind and the body is indeed a formidable task. It is because the mind is conceived as *res cogitans* manifested in the form of thinking, willing, imagining, experiencing, etc. which are not determined by laws of nature. The principle of causality as laws of nature puts a closure to the very explanation of the physical phenomena. The body as a physical phenomenon exists independent of the activities of the mind. Thus, Descartes also does not deny the existence of *res extensa*. As Churchland puts it, "Descartes theorized the reality divides into two basic kinds of substance. The first is ordinary matter, and the essential feature of this kind of substance is that it is extended in space: any instance of it has length, breadth, height, and occupies a determinate position in space. Descartes did not attempt to play down the importance of this type of matter. On the contrary, he was one of the most imaginative physicists of his time, and he was an enthusiastic advocate of what was then called "the mechanical philosophy". But there was one isolated corner of reality he thought could not be accounted for in terms of the mechanics of matter: the conscious reason of Man. This was his motive for proposing a second and radically different kind of substance, a substance that has no spatial extension or spatial position whatever, a substance whose essential feature is the activity of thinking. This view is known as *Cartesian Dualism*." (Churchland 1988: 8)

One of the major criticisms against Descartes' substance dualism has been expressed by Gilbert Ryle. Ryle an exponent of philosophical behaviourism argues that the mind and the body are not two independent substances. As Descartes' philosophical position on dualism, that is, 'the human body is in space like any other physical bodies and hence the bodily processes can be explained by using the mechanical laws. Such an explanation of the body is always a public state of affairs. In other words, the study of the body as a scientific study could be demonstrated. There is nothing private or secrete about body. On the other hand, the mind refers to thinking which is harnessed with the body. Thinking is attributed to the general human beings, rather than to any infant and insane. The notion of death and decay are ascribed to the body but not to the mind. The mind functions and continues to exist even after the death of the body.' For Ryle, this thesis of Descartes could be stated as an *official theory*. In brief, the official theory upholds the doctrine of substance dualism. The mind and the body are two independent substances and hence their existence can be conceived independently.

However, a person is combination of the body and the mind. The bodily changes do influence the mind and the *vice versa*. The changes in the body that occurs as the events occur in the external world. These events are observable facts. For instance, the change in my appearance, look, etc. can be stated just by studying certain biological factors. The changes that occur in the level of thoughts are personal. The person only knows what is going on in his inner mental life. A person has direct knowledge about the thoughts or mental states. What a person desires to do or intending to perform could be only known to him. A person as thinking mind evaluates the desires or could deliberate on the desire. The possibility of knowing directly and immediately is called *introspection* or *self-observation*. That is to know 'what is wanted or being felt.' Ryle writes, "In consciousness, self-consciousness and introspection he is directly and authentically apprised of the present states and operations of his mind. He may have great or small uncertainties about concurrent and adjacent episodes in the physical world, but he can have none about at least part of what is momentarily occupying his mind." (Ryle 2000: 14) He/ she are perfect authority to express the thoughts and feelings in the form of autobiography. It is because one has *direct access* to his or her own mind. The realm of the mental in this sense involves a sense of privacy. This has further given birth to the notion of *inner*.

Thinking about one's own feeling, thoughts, and experience are very much part of the inner which is not accessibly to the other. The other is very much part of the outer world. For example, I am only conscious of things that are going on in my mind; planning to pursue a career in academics, the experience of the beautiful sight of my village, pain of the loss of my father, etc. All these are part of *my* thinking, which people are not aware of, but if expressed to someone then he/she can write a biography of my personal life. In this regard, the very consciousness of one's own life brings in the mental existence of thoughts, feelings, experience, etc. There is physical existence as well which would refer to how I live as a bodily being. The living here is not dissociated from the body, rather logically separable for the very engagement that I have with the other and the world. For instance, I may not like my neighbor, but do not ever express this feeling to him. We continue to live as good neighbors. The feeling of mine is personal and very much part of my thinking that constitutes me as a person. The things happening in the outer world are necessarily having physical existence which is explained with reference to space and time. Whereas, the inner mental life of a person is not exhibited to anyone else, it is only given to *me* – to my consciousness.

Gilbert Ryle poses a serious doubt concerning the very existence of the inner. The inner as locus of the self is doubted seriously as it is not accessibly to the public/ the other. To assert on the existence of the self that resides in the inner mental life is dogma – that is, the 'dogma of the ghost in the machine.' One can offer a systematic explanation of the various parts of the machine – that is, how does a machine function? The machine represents a mechanical system constituted of various subsystems. Each subsystem is causally connected to one another as our bodily parts are assembled and constitute the whole organic system – the body. To say that the mind constitutes a separate category and is different from the body is a grave mistake. According to Ryle, this mistake is committed by Rene Descartes while separating the mind from the body and postulating the fact that they are categorically two different independent substances. While talking about the self, we generally refer to various voluntary actions, expressions or behaviour of the person which are indeed bodily, but to say that the mind is hidden inside these bodily expressions is to commit the mistake. What Ryle intends to suggest here is that the failure of providing an ostensive reference is not really a failure, because it happens in most of the cases such as the notion of University, team spirit, and the like?

For example, he writes if a foreigner visits to Oxford University, goes to different departments, library, gymnasium, laboratories, hostels, etc. and then asks where the university that is a rediculus question is. He also cannot say *pointing* to a department that 'this is Oxford University.' Thus it is important to note that what kind of terms are sought to have reference. Ryle writes, "My destructive purpose is to show that a family of radical category-mistakes is the source of the double-life theory. The representation of a person as a ghost mysteriously ensconced in a machine derives from this argument. Because, as is true, a person's thinking, feeling, and purposive doing cannot be described solely in the idioms of physics, chemistry and physiology, therefore they must be described in counterpart idioms. As the human body is a complex organized unit, so the human mind must be another complex organized unit, though one made of a different sort of stuff and with a different sort of structure. Or, again, as the human body, like any other parcel of matter, is field of causes and effects, so the mind must be another field of causes and effects, though not (Haven be praised) mechanical causes and effects." (Ryle 2000: 19-20) The acceptance of the mind and the body as two different substances is the cause of the category mistake. According to Ryle, these substances are not categorically independent from each other as it has been conceived by Descartes.

The mind as substance has only a *ghostly presence* in the entire organic system of the body. The bodily activities can be causally explained and so also the mental activities. The mind and the body are not separable entities. Since Descartes did not accept the so called mechanical theory which explained the movements of the body existing in space, he assumed that there could be a non-mechanical cause responsible for explaining the non-spatial activities. The mind has been entirely conceived as a different sort of things. Along with this point of view, Ryle assumed another factor which contributed to 'category-mistake' is the notion of the moral. He writes, "Since, according to the doctrine, minds belong to the same category as bodies and since bodies are rigidly governed by mechanical laws, it seemed to many theorists to follow that minds must be similarly governed by rigid non-mechanical laws. The physical world is a deterministic system, so the mental world must be a deterministic system. Bodies cannot help the modifications that they undergo, so minds cannot help pursuing the careers fixed for them. *Responsibility*, *choice*, *merit* and *demerit* are therefore inapplicable concepts – unless the compromise solution is adopted of saying that the laws governing mental processes, unlike those governing physical processes, have the congenial attribute of being only rather rigid. The problem of Freedom of Will was the problem how to reconcile the hypothesis that minds are to be described in terms drawn from the categories of mechanics with the knowledge that higher grade human conduct is not of a piece with the behavoiur of machines."

(Ryle 2000: 21-22) Descartes argument in favour of the non-mechanical explanation of human mind is something that retains the problem of dualism. Science accepts that law of causation can explain everything existing in nature. Thus determinism holds an extreme point of view that everything is causally determined. Ryle maintains that unless we leave the language in which the two substances are explained, dualism cannot be overcome. According to him, the mental and the physical belong to one category. And, we need to formulate expression that conjoins the mind with the body. He writes, "When two terms belong to same category, it is proper to construct conjunctive propositions embodying them. Thus a purchaser may say that he bought a left-hand glove and a right-hand glove, but not that he bought a left-hand glove, a right-hand glove and a pair of gloves." (Ryle 2000: 23) This is analogous to the notion that there exist no two processes: the mental process and physical process, rather they belong to one category of processes. To emphasize that there are only physical processes that cause mental actives and behaviour is to support the theoretical points of views of materialism. On the other hand, to say that mind is real and the bodily activities are known and construed through the mind is to adhere to idealism. In some sense, Ryle shows, if these two theses are uphold then we would commit *reductionism*.

(Ryle 2000: 24) If this is true then to accept one among the two is not a simple method of outing from the framework of dualism, as it has been thought by Churchland. (Churchland 1998: 18) The materialistic approach can only show us 'the explanatory advantage' of neuroscience, so far as the various complex mental activities are concerned. It has been known that brain is central to all the neurophysiological activities and its relevance is found in the area of medicine and brain pathology. (Churchland 1998: 19) Showing the strength of the argument of dualism, Churchland writes, "The dualist can admit that the brain plays a major role in the administration of both perception and behaviour – on his view the brain is the mediator between the mind and the body – but he may attempt to argue that the materialist's current successes and future explanatory prospects concern only the meditative functions of the brain, not the central capacities of the nonphysical mind, capacities such as reason, emotion, and consciousness itself. On these latter topics, he may argue, both dualism and materialism currently draw a blank." (Churchland 1998: 19)

## The Concept of Mind

The Concept of Mind is a 1949 book by philosopher Gilbert Ryle, in which Ryle argues that "mind" is "a philosophical illusion hailing chiefly from René Descartes and sustained by logical errors and 'category mistakes' which have become habitual." The work has been cited as having "put the final nail in the coffin of Cartesian dualism" and has been seen as a founding document in the philosophy of mind, which received professional recognition as a distinct and important branch of philosophy only after 1950.

## Critique of Cartesian Dualism

Ryle rejects Descartes' theory of the relation between mind and body, on the grounds that it approaches the investigation of mental processes as if they could be isolated from physical processes. In order to demonstrate how this theory may be misleading, he explains that knowing how to perform an act skillfully may not only be a matter of being able to reason practically but may also be a matter of being able to put practical reasoning into action. Practical actions may not necessarily be produced by highly theoretical reasoning or by complex sequences of intellectual operations. The meaning of actions may not be explained by making inferences about hidden mental processes, but it may be explained by examining the rules that govern those actions.

According to Ryle, mental processes are merely intelligent acts. There are no mental processes that are distinct from intelligent acts. The operations of the mind are not merely represented by intelligent acts, they are the same as those intelligent acts. Thus, acts of learning, remembering, imagining, knowing, or willing are not merely clues to hidden mental processes or to complex sequences of intellectual operations, they are the way in which those mental processes or intellectual operations are defined. Logical propositions are not merely clues to modes of reasoning, they are those modes of reasoning.

The rationalist theory that there is a *transformation into* physical acts of some purely mental faculty of "Will" or "Volition" is therefore a misconception because it mistakenly assumes that a mental act could be and is distinct from a physical act, or even that a mental *world* could be and is distinct from the physical world. This theory of the separability of mind and body is described by Ryle as "the dogma of the ghost in the machine". He explains that the workings of the mind as it governs the body are neither an independent nor a distinct mechanism, that there is no entity called "Mind" inside a mechanical apparatus called "the body", but that the workings of the mind may be better conceptualized as the actions of the body.

Cartesian theory holds that mental acts determine physical acts and that volitional acts of the body must be caused by volitional acts of the mind. This theory, according to Ryle, is "the myth of the ghost in the machine".

There *is* no contradiction between saying that an action is governed by physical laws and saying that the same action is governed by principles of reasoning. The motives of observable actions are *propensities* and *dispositions*; these explain why behaviors occur, and not some purely mental process. For example, the disposition to want or not to want something is not explained by an intellectual motive for that thing. The disposition to want something is explained by *the behaviors that are involved* in wanting that thing. Thus, the mind does consist of abilities and dispositions that do explain behaviors, for example the learning, remembering, knowing, feeling, or willing behaviors. However, personal abilities and dispositions are not the same as mental processes or events. To refer to abilities or dispositions as if they were purely mental occurrences is to make a basic kind of category-mistake.

The nature of a person's motives may be defined by the actions and reactions of that person in various circumstances or situations. The nature of a person's motives in a particular situation may not necessarily be determined by any hidden mental processes or intellectual acts within that person. Motives may be revealed or explained by a person's behavior in a situation.

Ryle criticizes the theory that the mind is a place where mental images are apprehended, perceived, or remembered. Sensations, thoughts, and feelings do not belong to a mental world which is distinct from the physical world. Knowledge, memory, imagination, and other abilities or dispositions do not reside "within" the mind as if the mind were a space in which these dispositions could be placed or located. Furthermore, dispositions are not the same as behavioral actions, but actions may be explained by dispositions.

Dispositions are neither visible nor hidden, because they are not in the same logical category as behavioral actions. Dispositions are not mental processes or intellectual acts, they are propensities which explain various modes of behavior. Perceptions, thoughts, emotions, and feelings may be understood as observable behaviors which have various modes of production.

Ryle admits that his approach to the theory of mind is behavioristic in being opposed to the theory that there are hidden mental processes that are distinct from observable behaviors. His approach is based on the view that actions such as thinking, remembering, feeling, and willing are revealed by modes of behavior or by dispositions to modes of behavior. At the same time, however, he criticizes both Cartesian theory and behaviorist theory for being overly mechanistic. While Cartesian theory may insist that hidden mental events produce the behavioral responses of the conscious individual, behaviorism may insist that stimulus-response mechanisms produce the behavioral responses of the conscious individual. Ryle concludes that both Cartesian theory and behaviorist theory may be too rigid and mechanistic to provide us with an adequate understanding of the concept of mind.

## Category Mistakes

As a linguistic philosopher, a significant portion of Ryle's argument is devoted to analyzing what he perceives as philosophical errors based in conceptual use of language.

His critique of Cartesian dualism refers to it as a "category mistake." Category mistakes such as the ones Ryle makes note of are made by people who do not know how to properly wield the concepts with which they are working. Their puzzles arise from the inability to use certain items in human language. A much-cited example is of a foreign visitor being shown round Oxford (which has no campus) and after having been shown colleges, libraries, laboratories and playing fields asks in puzzlement "But where is the university?" The answer is, of course, all of these.

The theoretically interesting category mistakes are those made by people who are perfectly competent to apply concepts, at least in the situations with which they are familiar, but are still liable in their abstract thinking to relocate those concepts to logical types to which they do not belong.

The dualist doctrine establishes a polar opposition between mind and body. At the language level, the mental properties are logical negations (in the Aristotelian sense) of the physical properties. So they belong, in accordance with the concept of category, to the same logical types, given that the expressions that are used for the descriptions of mental events are always mere negatives of the expressions used for the descriptions of material events. Ryle then says that such use implies a 'categorical mistake' for the descriptions of mental events that do not properly belong to the categories used for describing the corporeal events. Hence, 'mind' and 'matter' cannot be the polar opposites that Dualism suggests. Ryle writes that this would be comparable to claiming that "She came home in floods of tears" and "She came home in a sedan chair" (from the sentence "Miss Bolo came home in a flood of tears and a sedan chair", a zeugmatic sentence from Dickens) to be polar opposites. Such mistakes are, from the Rylean standpoint, the dogma of the mental ghost in the corporeal machine. Then, dualist doctrines are mythic in an analytical sense.

## Antecedents

Ryle builds on the work of the philosophers Ludwig Wittgenstein and Arthur Schopenhauer, among others. According to Bryan Magee, the central thesis of *The Concept of Mind* and the essentials of its subsidiary theses were derived from Schopenhauer, whose works Ryle had read as a student, then largely forgotten. Ryle, who believed that he had expounded an original theory, did not realize what he had done until someone pointed it out to him after the book was published.

## Reception

*The Concept of Mind* has been compared to Jean-Paul Sartre's *Being and Nothingness* (1943) by Iris Murdoch, who writes that English analytic philosophy shares the same general orientation as continental philosophy. Ryle has been interpreted by David Stannard as maintaining that the psychoanalytic idea of the unconscious is rooted in the Cartesian conception of a body-mind dichotomy and as such is one version of the "Ghost in the Machine" fallacy. According to Stannard, Ryle views the dogma as a logical error based on a category mistake.

Author Richard Webster, in *Why Freud Was Wrong* (1995), praises Ryle's clarity and strength of argument, but suggests that while his arguments effectively dissolve the mind-body problem, they have failed to bring about a revolution in human knowledge. Webster attributes this to the fact that Ryle's case that subjective aspects of experience such as sensation, memory, consciousness and sense of self are not the essence of "mind" has not been universally accepted by contemporary philosophers, neuroscientists, and psychologists. Webster believes that Ryle's willingness to accept the characterization of *The Concept of Mind* as behaviorist misrepresents its more nuanced position, writing that Ryle's acceptance of that description is not harmless, as Ryle himself suggested. Webster stresses that Ryle does not deny the reality of what are often called internal sensations and thoughts, but simply rejects the idea that they belong to a realm logically distinct from and independent of the external realm of ordinary human behaviour.

Ryle has been characterized as an "ordinary language" philosopher. The book's style of writing has attracted comment. Stuart Hampshire remarked in a review in *Mind* that, "There is only one property which I can discover to be common to Professor Ryle and Immanuel Kant; in both cases the style is the philosopher - as Kant thought and wrote in dichotomies, Professor Ryle writes in epigrams. There are many passages in which the argument simply consists of a succession of epigrams, which do indeed effectively explode on impact, shattering conventional trains of thought, but which, like most epigrams, leave behind among the debris in the reader's mind a trail of timid doubts and qualifications."

Its style of writing was commented on more negatively by Herbert Marcuse, who observes that the way in which Ryle follows his presentation of "Descartes' Myth" as the "official doctrine" about the relation between body and mind with a preliminary demonstration of its "absurdity" which evokes "John Doe, Richard Roe, and what they think about the 'Average Taxpayer'" shows a style that moves "between the two poles of pontificating authority and easy-going chumminess", something Marcuse finds to be characteristic of philosophical behaviorism. John Searle, who believes that no great work of philosophy contains many footnotes and that philosophical quality varies inversely with the number of bibliographical references, considers the absence of footnotes in *The Concept of Mind* a sign of its quality.

## Property Dualism

Searle is often being interpreted as Property Dualist. As he holds that there are mental properties to describe the nature of mental states and there are physical properties to describe the nature of neurophysiological states. To say that the mental states are having certain specific properties, is to maintain that they are conscious, possess intentionality, they can cause action, and moreover, they are of person's mental life. In brief, these features are called; consciousness, intentionality, mental causation, subjectivity, etc. The mental properties, according to Searle, are real and intrinsic to consciousness. They are intrinsic to consciousness because consciousness has been treated as the most fundamental feature mind. These are intrinsic to the mind as they constitute consciousness. If consciousness is reduced to certain physical process, consciousness becomes external to the brain. In that case they become second order properties. Therefore, they cannot have first order intentionality. Now, one may also ask, does their irreducibility result in property dualism? Or, if they are so intrinsic to the consciousness do they lead to substantial dualism? Searle's biological naturalism rejects any form of dualism. But it does talk about naïve mentalism and naïve physicalism. Both mental and physical are real and co-exist in nature. The mental is the part of the physical and thus consciousness, along with its co-features are caused by the brain processes.

As Searle says, "One way to characterize this position is to see it as an assertion of both physicalism and mentalism. Suppose we define 'naïve physicalism' to be the view that all that exists in the world are physical particles with their properties and relations. ...'naïve mentalism' to be the view that mental phenomena really exist. They really are mental states; some of them are conscious, many have intentionality; they all have subjectivity, and many of them function causally in determining physical events in the world... Naïve mentalism and naïve physicalism are perfectly consistent with each other. Indeed, as far as we know anything about how world works, they are not only consistent, they are true."[1] Thus Searle does not deny the dual nature of the reality; it being physical and mental at the same time naïve mental ism, however is not a strong form of dualism. It does not solve the problem of Cartesian dualism. To say that mental is realized in the brain process is to say that mental is dependent on the physical. On the other hand, the irreducibility of the mental leads to ontology of consciousness. The conscious mental phenomena are the emergent properties of the neurological processes of the brain and so cannot be identical with the brain processes.

Searle writes, "The existence of consciousness can be explained by the causal interaction between the elements of the brain at the micro-level, but consciousness cannot itself be deduced or calculated from the sheer physical structure of the neurons without some additional account of causal relation between them....If consciousness were emergent, then consciousness could cause things that could not be explained by the causal behaviour of the neurons. The naïve idea here is that consciousness gets squirted out by the behaviour of the neurons in the brain, but once it has been squirted out then it has a life of its own." We have been repeatedly insisting on this point in our earlier sections that according to Searle, consciousness has its independent function though it is caused by the neural processes. Therefore, for him, the cause-effect relationship is significant. Searle is aware of the danger that causal reductionism will result in the ontological reductionism. However, Searle is in a dilemma: either biological naturalism will fall in the same stream of reduction or his defense of first person ontology will result in some sort of Cartesian dualism. Searle is opposed to both.

However, the claim of Searle is that the perfect scientific explanation of the mind does not lead to ontological reductionism. Causality is not rejected fully by the scientific explanation of the mind. Rather the analysis that he tries to project clubs both scientific and pre-scientific experience with of phenomena. The pre-scientific experience here refers to the common experience of things in the world. Our experience of water as liquid that quenches thirst is not scientific explanation. The real scientific explanation tell us what causes water as a liquid, i.e. its combination of two hydrogen molecules and one oxygen molecule. In that case can we discard the pre-scientific understanding of the water. Similarly, in the case of pain, there is a physiological process involved in it. And the mental aspectuality is also not denied. By aspectuality here we mean the subjective experiences of pain or the first person feeling of pain. The question before us is not only to include the real causal analysis or scientific analysis of the pain behaviour or experiences of water etc., but to include our pre-scientific experiences with the phenomena as well. And this can be applied to the notion of consciousness. It has both pre-scientific as well as scientific explanation. For Searle, "Part of the point of reduction was to curve off the subjective experiences and exclude them from the redefinition of the real phenomena, which are now defined in terms of the features that interest us most. But where the phenomena that interest us most are the subjective experiences themselves, there is no way to curve anything off ...indeed it is a general feature of such reductions that the phenomenon is defined in terms of "reality" and not in terms of "appearance". But we cannot make that sort of appearance-reality distinction of consciousness because consciousness consists in appearances themselves. *Where appearances are concerned we cannot make the appearance reality distinction because appearance is the reality.*" The appearance of consciousness is the various conscious activities. These are the macro level function consciousness. All our macro level activities carry features consciousness and other properties of it. On the other hand, consciousness refers to the macro-level physical reality of the brain because it is caused by the neurological structure of the brain organism. This, however, is not the appearance-reality distinction. In the case of consciousness, no such distinction arises.

Dennett questions the Searlean way of analyzing the mental states and the properties. He questions the *ontological status* of the mental phenomenon. As he says, "When are elements of the pattern real and merely apparent? Answering this question will help us resolve the misconceptions that have led to the proliferation of 'ontological positions' about beliefs, the different grades or kinds of realism." To get into the ontological status of the mental states the method of study must be *objective*. Once we understand the designing pattern of the organism or the system as assembled of, we could then well detect the phenomena and their intrinsic value. The objective features of the mental states are referred to by the micro level structure of the brain organism. Because that is the root and the center of controlling the behaviour as far as Dennett's analysis is concerned. Dennett says: "We can adopt the physical stance toward a configuration of Life world our powers of predictions are perfect: there is no noise, no uncertainty, no probability less than one. Moreover, there are no back stage hidden variables; the unfolding of the physics of life world is directly and completely visible."

The physical stance signifies the neurological structure of the brain organism causing neural states and operating them further to cause different behviour. Dennett's sub-system analysis explains clearly the operations of the sub-organism involved in the function of neural or brain states. Brain as a biological organism is part of the life world and is accessible to the 'physics of life'. The entire physics of life world is such that it is completely explained by the 'single unexceptional law, which .is the 'foundational law of physics.' The physical laws could explain the various features that occur in the biological system. The real pattern that exists in reality must be amenable to description by the physical laws. However, it is difficult to locate the 'intentional order' in the life world. He writes: "If you "look at" the world in the right way, the patterns are obvious. If you look at (describe) the world in any other way, they are, in general, invisible." Furthermore, he maintains that "A pattern exists in some data¾is real¾if there is a description of the data then there is more efficient than the bit of map, whether or not alone can concoct it.

Dennett mentions that the Searlean model always insists upon the deeper structure analysis of the mental states but the search has been *forlorn*. Once we go into the deeper fact analysis we proceed through the structural configurations of the organisms for causing such features. And the consequence may prove that there is nothing called original intentionality or real intentionality. It is because, real intentionality not only has a subjective appearance but needs to have the micro-fact or deeper fact. Dennett's complain is that their supposition fails to *identify* such fact. Dennett's analogy of human behaviour and the behaviour of intelligent robot nullifies Searlean notion of intrinsic features. Rather their function shows that mental states carry *as if* intentionality. To say that the micro-level structure is real in the sense of causing the subjective appearances reveals the whole designing of the system or the organism as purely unconscious stuff. And the designed is the source. Then this notion of original intentionality is also a phenomenon derived directly from the body structure and indirectly from the whole of natural selection.

However, Searle treats the intentional properties as real and basic to the very notion of our conscious activities. The conscious activities of human beings reveal the structure of consciousness. The structure of consciousness further reveals its constituents as intrinsic to the very notion of consciousness. Since our *subjective* attitude is a part of the constitutive features of consciousness, it has been treated as an irreducible ontological feature. Searle tries to naturalize this *subjective ontology* in his theory of biological naturalism. That is, he treats the subjective attitude as part of the biological capacity of the neural processes of the brain. It has a neurological basis. Thus the subjective ontology has a special status with reference to its *independent* function throughout its *intentional* interaction with the world. The intentional interaction is basic to the conscious participation or interaction because it is the cause of bringing out further intentional states and *realization* of those emerging states. Since Searle's emphasis is always on the commonsensical understanding of the mind, he is not able to deny the subjectivity as the prime feature of consciousness. It is because, everybody has his or her way of looking at the phenomena and their experience differing accordingly. And insofar as the ontology of subjectivity is concerned it remains always the first-person ontology. My awareness of my mental states and attitudes is certainly distinct from that of others as their attitudes differ from mine.

As Searle says, "This subjectivity is marked by such fact that I can feel my pains, and you cannot. I see the world from my point of view; you see it from your point of view. I am aware of myself and my internal mental states as quite distinct from the selves and mental states of other people." Thus for him, subjectivity is an objective phenomenon. Dennett questions this notion of Searle's "subjective ontology." He writes, "it is highly idiosyncratic: what is special about consciousness is it subjective ontology alongside (not within) otherwise science can accommodate subjective ontology alongside (not within) otherwise objective ontology. Once we clear away some widespread confusions about what science requires, and dismisses the misbegotten fluid of cognitive science that has been engendered by those confusions the subjective ontology of the mind, be claims, will loss its aura of unacceptable mystery." Since Dennett's methodology is completely *hetrophenomenological* he denies the very notion of *central meaner* or the subjectivity as the core point of seeing and experiencing the phenomena. By positing subjectivity as a central or irreducible factor Searle brings in Cartesianism. But the Cartesian subjectivity is the notion that both Searle and Dennett reject. Dennett's heterophenomenological analysis explains away the very notion of consciousness and thereby the notion of subjectivity.

Searle's aim in rejecting the Cartesian subjectivity is to naturalize the very concept of subjectivity in the intentional framework of biological naturalism. That is the reason, Searle posits, a specific notion of mental causation which retains subjectivity. According to Searle, there are three mistakes the Cartesians commit. They are,

1. "All consciousness are self conscious.
2. Consciousness is known by the special feature of introspection.
3. Knowledge of our conscious states are incorrigible."

For Searle, this sort of reflexive consciousness cannot be denied. He says "In short, if by 'introspection' we mean simply thinking about our own mental states, then there is no objection to introspection. It happens all the time, and is crucial for any form of knowledge. But if by introspection we mean, special capacity, just vision only less colourful, that we have to *spect intro* then it seems to me there is no such capacity. There could not be model of specting intro requires the distinction between the object *spected* and the *specting* of it, and we cannot make this distinction of conscious states." The Cartesian notion of 'introspection' refers to the analysis of thought and thought processes as an exclusive feature of mental activity. Therefore in the *Meditations* Descartes emphasizes the *reflexive act* of the thought processes. It is the reflexive feature of consciousness which is a property of thinking. Moreover, reflexivity is a kind of *thinking about thinking*. As Descartes while clarifying the unique feature of consciousness says, "...it is not sufficient for it to think: it is further required that it should think that it is thinking, by means of a reflexive act, or that it should have awareness of its own thought."

Reflexivity as a feature of consciousness is natural to the very nature of consciousness. One need not always reflect on this skill in order to establish it. He says, "Reflexivity is not required in order for a thinking substance to be superior to matter. The initial thought by means of which we became aware of something does not differ from the second thought by means of which we become aware that we are aware of it, any more than this second thought differ from third thought by means of which we become aware that we are aware that we were are aware."

Descartes in this basic sense does not differ from Searle. The disagreement is not on whether there self consciousness or not, but on the fact that it is another *special faulty* of consciousness. As Descartes mentions, "we can direct our mental state at another; we can think about our thoughts and feelings; and we can have feelings about our thoughts; but none of these involves a special faculty of introspection." (op.cit.) Reflexivity in Descartes however, is not attached with a special faculty; rather it is the special feature of consciousness. As he maintains, "But it must be noted that, although we are always actually aware of the acts or operations of our mind, we are not always aware of minds faculties or powers, except potentially. By this, I mean that when we concentrate on the employment is of our faculties then immediately, if the faculty in question resides in our mind, we become actually aware of it, hence we may deny that it is in the mind if we are not capable of being aware of it." This self-awareness is about the inner capacities of the mental life. Hence it has been treated as special feature of intentionality of consciousness. For Searle, the change of intentional reference implies the *shifting* of consciousness. Shifting is special feature of consciousness in having inner directedness. In this regard Searle defends the self-awareness.

And all our conscious activities are the part of this self-consciousness. All the conscious representations together constitute the notion of self-consciousness. Searle is not definitely talking about discrete consciousness. Rather consciousness for him is having *unity*. Searle talks about two sorts of unity the vertical and horizontal unity. The vertical unity describes the assembling features of all conscious experiences. As Searle writes, "Vertical unity is matter of simultaneous awareness of all diverse features of any conscious states ...we have little understanding how brain achieves this unity. In neurophysiology it is called "binding problem" and Kant calls the same phenomenon the transcendental unity of perception." This simultaneous awareness is a part of the subject's consciousness activity. And our ordinary notion of conscious experience and the reflexive consciousness are related to the notion of self-consciousness. It is because they all together constitute self-consciousness. Consciousness is the foundation of this self-consciousness. Searle does not deny fully the notion of self consciousness. He remarks, "To know consciousness with certainty, we must know it by a special faculty that gives us direct access to it. Hence, the doctrine of introspection is relevant. And though I am less confident about historical diagnosis ¾ if self is the source of all knowledge and meaning, and therefore to be based on its own consciousness, then it is natural to think that there is a necessary connection between consciousness and self-consciousness; hence the doctrine of self-consciousness." Thus, the subjective nature of consciousness entails the self consciousness. Searle accepts unity of consciousness which is unfolded in self-consciousness. Hence, self-consciousness is the root of these underlying features of all our conscious activities. As he says, "It is indeed the case that conscious mental states have subjective ontology" In this regard the mistakes that Searle finds in the Cartesian tradition do not seem to be a mistake at all.

## The Concept of Materialism

Materialism argues against the above Cartesian conception of mind. The Cartesian conception of the reality divides the mind and the body into two different realms of existence. And insofar as consciousness is concerned, it is the metaphysical foundation for the scientific as well as commonsensical explanation of the mental phenomena. Materialism as an alternative conception of mind explains the reality of the mental through the reality of the physical. The concept of body is defined inter terms extension. The body exists as an independent substance. Its existence does not depend on the existence of the mind. Thomas Hobbes has initiated the study of the mechanical function of the body. According to him "The body is also called the 'subject', because it is so placed in and subjected to imaginary space, that it may be understood by reason, perceived by senses." The body as a *subject* for the discourse of experience and action necessitates a separate ontological status. In this regard, Hobbes was interested in studying the causal function of the body. The function of the body is an *accident* – 'an accident is that faculty of any body by which it works in us a conception of itself.' For him, the conception of the body is such that it has the power to *bring out* certain effects. So the production of effect presupposes the causal power of the body. 'If body *A* generates motion in body *B*, *A* is the agent and *B* is the patient. Thus, if fire warms my hand then fire is the agent and the hand is the patient.'

The agent-patient relationship shows the active and passive power of the body so far as the change of events is concerned. So far as human beings are concerned, there are two kinds of events; the mental events and the physical events. For the materialist, the mental events are causally connected to the bodily/ physical events of the body. In other words, the mind is just a part of the physical organism. The function of the organic system would unfold the causal relationship existing in the system and causing action.

In this regard, the mind can be known through the various activities of the body/ the brain, i.e., it involves the physical behaviour of the brain comprising of the different states and events that are occurring in the brain. Hence there is no dualism between the mental and the physical. Materialism includes various theories of the mind like behaviourism, physicalism and functionalism. All these theories talk about some sort of identity relationship between the mental phenomena and the neurophysiological states and processes of the brain.

Behaviourism, as a theory of mind, advocates the empirical study of the mind through behaviour. As Churchland puts it, "By 'behaviour', behaviourists' mean the publicly observable, measureable, recordable activity of the subjects at issue: bodily movements, noises emitted, temperature changes, chemical released, interaction with environment, and so forth." It is the behaviour which is the source of our knowledge of the mind. Rather mind, as such, is nothing but the behaviour itself. We do not, therefore, observe the mental phenomena. It is basically the behaviour which is observed. And this observable behaviour is to be taken into account in order to explain the mental phenomena.

In this sense, the definition of the behaviour of human beings includes both voluntary or conscious actions as well as involuntary or unconscious actions. Voluntary action includes the person's intention and will, whereas the involuntary actions exclude the persons' intention and desires. It includes the biological actions of the body. The behaviour is studied by undertaking both the outer environmental conditions as well as the inner potential capacities of human beings. As Churchland puts it, "Instead of appealing to mental states, behaviourists proposed to explain any organism's behaviour in terms of its peculiar environmental circumstance. Or, in terms of environment plus certain observable features of the organism. Or, in terms of certain *un*observable features of the organism¾ dispositions, and innate and conditioned features meet a very strict condition: they must be such that their presence or absence could always be decisively determined by behavioural test, as solubility of sugar is revealed by its actually dissolving (the behaviour) when placed in water (the environmental circumstances)." Insofar as the mental state is concerned, there are two aspects which count mostly for explaining the mind: they are the dispositional capacity as an intrinsic feature of the human beings' organic structure and the environmental condition as an outer condition in which a person is being situated. Environment is an external condition which has a direct impact on human behaviour.

Behaviourism as a philosophical thesis reasonably motivated to critique dualism. Gilbert Ryle's response to Descartes' *substance dualism* is a famous example of philosophical behaviourism. Ryle's philosophical behaviourism is stirred by the development of philosophical thesis of logical positivism and the 'general assumption of analytic philosophy that philosophical problems arise due to linguistic or conceptual confusions which can be dissolved by a careful analysis of language. The logical positivists emphasized that 'meaning of the sentences is determined by the observable circumstances.' Human behaviours are external observable facts. As the expressions of the mind, behaviours are public state of affairs. Descartes' conceptualization of the mind is a private phenomenon and does not exist in space. And it can be known through *introspection* was something problematic for Ryle. According to Ryle, the very division that Descartes draws between the mind and the body is the reason for creating substance dualism. Descartes commits a kind of *category mistake* by conceptually categorizing the mind and the body as two different substances. To place mind as a substance within the framework of the body is to conceptualize the existence of the mind in a dogmatic way. Ryle calls it '*the dogma of the ghost in the machine.*'

The body is conceptualized as an organic system that causally operates like a machine; to say that the mind or the soul resides in the centre of *pineal gland* and controls the bodily movements while performing voluntary actions, but the intervention of the mind not an observable fact. Rather, it is *directly* and *immediately* accessible itself. In other words, the soul is self-conscious – it is conscious and is *aware of* its existence. To explain consciousness with the help of consciousness is to commit the fallacy of circularity. The Cartesian explanation of the mind is thus a problematic thesis does not go along with the scientific discovery of the fact that there is a causal-mechanical process through which body operates. The behaviourists give importance to the study of actual and the potential pattern of the behavior. Emotion, belief, desire are not about 'ghostly inner episodes', rather they are observable when expressed in behaviours. For instance, desire to feel thirsty is a potential dispositional state of the living body. To quench the thirst, when someone looks for water or any kind of drinks implies state of belief. Ryle and other behaviourists talk about *multi-track dispositions*.

The inner feature of behavioural capacity is known as the dispositional property of behaviour. The *disposition* is an unobservable factor which gets manifested in the behaviour itself. For example, brittleness is the dispositional property of glassware. Glass, being a physical phenomenon, contains brittleness as its physical dispositional property. Similarly, human beings' behaviour is determined basically by the dispositional feature of the brain or brain states. That is, for all of our outer physical behaviour there must be a causal dispositional state in the brain. Dispositions are thus built into the structure of an object. For example, whenever any glassware falls from my hand or is hit heavily, it lies broken on the ground. The brittleness is a part of the atomic structure of each piece of the broken glass. It is also true that there is some external factor operating for causing the glass to slip from my grip. The disposition of the glass is simply that if breaks whenever it falls to the ground. Similarly, in certain environmental and physical conditions human beings behave in a particular order. Hence, in order to determine the human dispositions we must observe or measure the particular order of the external behaviour of the human beings. Besides in the case of human behaviour, we must observe the dispositional features of the brain states. Brain states, for the behaviourists, are the physical states. Behaviour is identified with the brain states, i.e. corresponding to a particular behaviour there must be a dispositional state in the brain organism. We generalize about dispositions by seeing the constant occurrence of the same type of behaviour in typical external condition. Thus we arrive at general causal laws regarding behaviour.

The mental phenomena like desire, belief, sensation, feelings, imagination, etc. are a part of the dispositional features of the brain, the brain being a physical organism that is directly causing behaviour. There is no further necessity for admitting the mental states and features as the cause of the behaviour. The reality of the mental is basically a physical dispositional feature of the brain. Thus behaviourism in a way rules out the reality of the mental as a necessary condition for explaining behaviour. Moreover, it only looks for the behaviour as such, as real, along with the physical conditions that are causing the behaviour.

Behaviourism as a theory of mind is strongly criticized by another group of materialist philosophers like J.C.C. Smart, D.M. Armstrong and U.T. Place. They reformulated the behaviourists' conception of identity between the dispositional states of the brain and the behaviour by introducing the hypothesis of *causal brain process* for the explanation of behaviour. The causal process of behaviour must have a scientific explanation. Scientific explanation deals with the law of causality. Since mental process is understood as a physical processes of the brain, the mental life needs a causal explanation. In that sense, the brain process can be taken as the cause of the mental phenomena.

D. M. Armstrong defends the central-state identity theory according to which the brain states are the same as the mental states. Each state of the mind is causally responsible for a piece of behaviour. Hence, there is a causal relationship between the states of mind and the expressions of behaviour. The behaviour is not simply an automatic process caused by the dispositional character of the mind. Rather the brain process is identical with the mental state.

Armstrong strives for a synthesis of two conflicting theories of mind, namely, the Cartesian theory of soul-substance which considers the autonomy of the mental as real and the behaviourists' theory which considers behaviour as the only reality. As he describes the synthesis, "My proposed synthesis is that the mind is properly conceived as an inner principle, but a principle that is identified in terms of the outward behaviour it is apt for bringing about. This way of looking at the mind and mental states does not itself entail a Materialist or Physicalist view of man, for nothing is said in this analysis about the intrinsic nature of these mental states. But if we have, as I have asserted that we do have, general scientific grounds for thinking that man is nothing but a physical mechanism, we can go on to argue that the mental states are in fact nothing but physical states of the central nervous system." The synthesis emphasizes the scientific study of the mental phenomena. The reality of the mental phenomena is established by identifying them with the central states of the nervous system. The central nervous system brings out the physical processes involved in the organism for causing several behaviour. And behaviour is causally explainable through the identification of the underlying physical processes of the brain. So neither behaviour nor the reality of the mental is ruled out. What the identity theory of the mind tries to look at is the inner principle which is the essential link between mind and behaviour.

This above conception known as the "Central-state Theory" involves two doctrines: the *psycho-chemical materialism* and the *attributive theory of mind.* Psycho-chemical materialism takes into account the processes of the nervous system of the brain and tries to formulate the physical and chemical laws for explaining the processes themselves. Armstrong writes, "[it]...is to identify these inner states with physical chemical states of the brain. This is a contingent and scientific identification, and it yields Central-state Materialism." Since the psycho-chemical materialism of Armstrong emphasizes the scientific study of the mental, it holds the contingent identity relationship between the mental states and the brain processes. This contingency is an observable fact about their identity relationship. And from the observable facts of the physico-chemical processes of the brain, psycho-chemical laws are formulated.

The attributive theory holds that mind is a causal attribution to the so called process of stimulus and responses. As Armstrong points out, "Man as an object is continually acted upon by certain physical stimuli. These stimuli elicit from him certain behaviour, that is to say, a certain physical responses. In the causal chain between the stimulus and response, falls the mind. The mind is that which causally mediates our response to stimuli. Now central state theory wants to say that between the stimulus and response fall physical processes in the central nerves system, nothing else at all, not even something 'epiphenomenal'." The reality of the mental is not ruled out. The functions of the mental states are identified with the functions of the brain processes. Brain as the central part of the whole organism, involves inner processes. That holds a causal link between outer behaviour and the inner brain process. "Central-state Materialism holds that, when we are *aware* of our mental states, what we are aware of are mere physical states of the brain. But we are not certainly aware of mental state *as* states of the brain." We are not aware of these mental states as identical with the brain processes. But our awareness of something happening in the mind is always about the physical processes that are going on in the brain. The identity principle holds good because only the physical processes of the brain are going on. Armstrong further puts it, "...the Materialists theory allows a peculiarly simple theory of the nature of unconscious mental states. For the materialist will say that, as a matter of empirical fact, they are simply a physical state of the brain. Now most modern philosophers would admit that in such a case as the patient struggling under nitrous oxide, or the chicken sexer sexing a chick, there are brain processes going on which are responsible for, or at any rate closely connected with this behaviour. And if we are forced to admit the possibility of unconscious mental states any way, it will be an aggregate economy to say that they are nothing but brain processes." Thus for any sort of action, there is a corresponding process in the neurophysiological structure of the brain. Actions are identified with those physical processes of the brain which are causing such behaviour.

U. T. Place finds Armstrong's identity thesis inadequate. It does not resolve the gap between the brain processes and the mental states. He extends the identity hypothesis further by advocating a strong identity between consciousness and the brain processes. He offers the hypothesis that consciousness is an outcome of the brain processes. According to Place, this hypothesis remains powerful till today because of its consistency with the empirical evidences. And thereby it has become an established thesis. However, Place's formulation of the *hypothesis* is such that the brain-mind identity is not a logical identity; it is an empirical theory based on scientific evidences. Therefore, for him, the statement " 'Consciousness is a brain process' is a reasonable scientific hypothesis. The hypothesis says that conscious states and events can be identified with certain pattern of brain activity. As he says "What do I want to assert, however, is that the statement 'consciousness is a process in the brain', although not necessarily true is not necessarily false. 'Consciousness is a process in the brain' in my view, is neither self-contradictory nor self-evident; it is a reasonable scientific, hypothesis, in the way, 'lightening is a motion of electric charges' is a reasonable scientific hypothesis." Consciousness *is* identical with brain processes. Here, 'is' has two functions; one is of *composition,* and other is of *definition*. Consciousness or conscious mental states are composed of certain physicochemical processes of the brain. And when we are consciously behaving, we can identify consciousness with behaviour, but always a physicochemical process does take place in the brain. Empirically, the conscious states are the physic-chemical processes in the brain. For instance, the cloud that we observe in the sky is nothing but a mass of rain-bearing particles. But, scientifically speaking, "A cloud is large semi-transparent mass with fleecy texture suspended in the atmosphere, whose subject is continual and kaleicdospic change". This is, however, not a logical definition of 'cloud'. So it is not a contradiction to say that cloud is a mass of rain-bearing particles.

According to Place, the identity relationship between consciousness and the brain processes consists in that the occurrences of the conscious mental states, and the change corresponding in micro structure of brain processes are simultaneous events. These, are in fact no two events occurring; there is only one event, though it is described in two ways. For instance, as Place says, one and the same phenomenon can be simultaneously observed from two points of view¾one from the scientific point of view and other from the common man's point of view, as we have seen in the example of the cloud. Both stand points are systematically and causally related with each other. One goes parallel with the other. Similarly, whenever we talk about consciousness we talk usually about the various conscious experiences, not about the micro physicochemical events and processes that go on in the brain. Thus our description of the mental phenomena is given through the secondary qualities that are attributed to the phenomenal field. This sort of description results in *phenomenological fallacy,* according to Place; because it ignores the fact that underlying our so called phenomenological experience, there is a physical process going on. Place writes, "We describe our conscious experience not in terms of the mythological "phenomenal properties" which are supposed to be inhere in the mythological 'objects' in the mythological phenomenal 'field', but reference to the actual properties of the concrete physical objects, events and processes which normally, though not perhaps in present instance, gives rise to the sort of conscious experience which we are trying to describe." So, for Place, the real description of our conscious experience must include the physical properties of the brain states and process. This is because; the brain process is the only reality that exists. And brain process is considered real, because it is the cause of our phenomenal experiences.

J.C.C. Smart advocates *strict* identity between the mental states and the brain processes. He supports Place' the notion of identity, because it emphasizes only the scientific aspects of notion of identity. It is not necessary to talk about the logical aspects of the identity. For example, lightening is scientifically described as electrical discharge. Scientific explanation itself is based on the contingent description of the lightening because it is also a logical possibility that this description of lightning may change after some other new scientific discovery about lightning. Smart says, "On the brain processes thesis, the identity between the brain process and the experience is a contingent one. So it is logically possible that there should be no brain process, and so processes of any other sort (no heart process, no kidney process, no liver process) There would be the experience but no physiological process with which we might be able to identify it empirically." Smart argues that though this is a possibility, empirically it is not the case. He asserts the relation of identity stating that all experiences are composed of the physical properties of the brain processes. This could be treated as an *a priori* notion because "what could be composed of nothing cannot be composed of anything".

In this regard the brain possesses physicochemical properties which cause sensation, feelings and other mental states. A conscious experience is a very complex process involving a vast number of neurons. It is a process, not stuff. Brain process is an inner phenomenon which needs to be identified by the advanced science of neurophysiology. He describes the complexity involved in the human mental life by saying: "... for a full description of what is going on in man you would have to mention not only the physical processes in the tissue, glands, and nervous system, and so forth, but also his states of consciousness: his visual, auditory, and tactual sensations, his aches and pains, etc. That these should be correlated with brain processes does not help, for to say that they are correlated is to say that they are something 'over and above'." Brain processes are not 'over and above' of all our conscious experiences including our linguistic behaviour. For example, when someone says he is experiencing pain in his right toe, he is experiencing the pain because there is stimulation of the C-fiber of the neurological structure of the brain processes. He says, "All it claims is, that in so far as a sensation statement is a report of something, that something is in fact a brain process, sensations are nothing over and above brain processes." Though the brain process is not publicly observable, it is an underlying factor responsible for all our mental activities. Moreover, Smart claims that "All that I am saying is that 'experience' and 'brain processes' may in fact refer to the same thing, and if so we may easily adopt a covention (which is not to change our present rules for the use of experience words but an addition to them) where it would make sense to talk of an experience in terms appropriate to physical processes."

## References


- Deuraseh, Nurdeen; Abu Talib, Mansor (2005). "Mental health in Islamic medical tradition". The International Medical Journal. 4 (2): 76–79
- Bishop, Paul (2000). Synchronicity and Intellectual Intuition in Kant, Swedenborg, and Jung. USA: The Edwin Mellen Press. pp. 262–267. ISBN 0-7734-7593-1
- "Soul (noun) - Oxford English Dictional (online full edition)". Oxford English Dictionary (OED). Oxford English Dictional (OED). Retrieved 1 December 2016
- Bahá'u'lláh (1976). Gleanings from the Writings of Bahá'u'lláh. Wilmette, Illinois, USA: Bahá'í Publishing Trust. pp. 158–163. ISBN 0-87743-187-6. Retrieved 2016-02-23
- Jones, David (2009). The Gift of Logos: Essays in Continental Philosophy. Cambridge Scholars Publishing. pp. 33–35. Retrieved 2016-02-23
- Smith, Joseph (1981). Doctrine and Covenants. Salt Lake City, Utah, United States of America: The Church of Jesus Christ of Latter-Day Saints. p. 286. ISBN 1592975038
- Shields, Christopher (2011). "Aristotle's Psychology (supplement: The Active Mind of De Anima iii 5)". The Stanford Encyclopedia of Philosophy. Retrieved 2013-12-12
- Creeger, Rudolf Steiner ; translated by Catherine E. (1994). Theosophy : an introduction to the spiritual processes in human life and in the cosmos (3rd ed.). Hudson, NY: Anthroposophic Press. pp. 42–46. ISBN 0-88010-373-6
- "Do Embryos Have Souls?", Father Tadeusz Pacholczyk, PhD, Catholic Education Resource Center". Catholiceducation.org. Retrieved 13 November 2011
- Barrette, Cyrille; Saint-Arnaud, Jean-Guy (2013). Lettres Ouvertes Correspondance entre un athée et un croyant (1 ed.). Québec: Mediapaul Canada. p. 130. ISBN 9782894209134. Retrieved 30 January 2016
- Farah, Martha J.; Murphy, Nancey (February 2009). "Neuroscience and the Soul". Science. 323 (5918): 1168. doi:10.1126/science.323.5918.1168a. Retrieved 20 November 2012

2

# Exploration of Cognitive Mind

The process of acquiring and gathering knowledge is known as cognition. It incorporates processes like memory, learning, attention, decision-making, etc. Some of the common experiments in the field of cognition are serial position, Brown-Peterson, memory span, visual search and word superiority. This chapter is an overview of the subject matter incorporating all the major aspects of cognition.

## Cognition

Cognition involves issues how we acquire, store, retrieve, and use knowledge. If we use cognition every time and acquire a bit of information, place it in, or use that information in some way, then cognition must include a wide range of mental processes. Cognition, generally, is a process in which information is encoded in the brain by receiving signals from the outer world through the sense organs. It seems that human cognition is largely different from that of other animals because of the enormous richness of the human cognitive process.

There are different models of understanding cognition such as the neuro-scientific, psychological, representational, computational model, and connectionist model. These five approaches explain cognition from different standpoints.

Neuroanatomy is the study of the nervous system's structure, and is concerned with identifying the parts of the nervous system and describing how the parts are connected to each other. Neuroanatomy can be made at many descriptive levels. According to neuroscientists, investigation can be made at two levels. Firstly, gross neuroanatomy is about general structures and connections, whereas fine neuroanatomy is the main task that describes the components of individual neurons. Secondly, histology is the study of tissue through dissection. The primary concern of neuroanatomy is to ideally connect the patterns of connectivity in the nervous system, and to layout the mechanism that allows information to get from one place to another. For neuroscientists, this neuroanatomy is the analogy of cognition.

While explaining neuro-scientific approach to cognition, Francis Crick, a famous neuro-scientist observes "the brain does not make a distinction between hardware and software, as a computer does. Theories [of thought] have made this distinction are unfortunate". Crick argues for theories of cognition that are strictly tied to biology, which implicitly force him to argue for the study of simple cognitive acts, such as visual word detection, rather than the complex acts such as paragraph comprehension, etc.

For neuroscientists, mental action depends on the psycho-neurological factors underlying in it. No mental action is without these factors which determine the mental history of an agent. Although Crick's general discussion is on the concept of conscious thought; virtually all the specific studies he cites deal with visual cognition. For him, it might be most profitable to deal with vision entirely within the field of neuro-science, while dealing with language comprehension in terms of psychological machines has no known neural basis. Cognition is a multi-dimensional process which needs a many-sided approach.

The psychological approach to cognition is defined as the psychology of understanding and knowing. It is also the study of mental processes. It is concerned with the way we take in information from the outside world, and how we make sense of that information, and what use we make of it. According to Groome, there have been three main approaches to the study of cognitive psychology, namely, experimental psychology, computer modelling, and cognitive neuropsychology. Firstly, experimental psychology involves the use of psychological experiments on human subjects to investigate the ways in which we perceive, learn, remember or think. Secondly, cognitive psychology is the use of computer modelling of cognitive processes. This approach involves the simulation of some aspects of human cognitive function by writing computer programs, in order to test out models of possible brain function. Lastly, cognitive neuropsychology is concerned with the activities of the human brain during cognitive processing. These three approaches of cognitive psychology explain the psychological levels of cognitions.

The representational theory of cognition tries to show how our knowledge of the world is represented in the mind. When human knowledge is represented in an abstract format, we call it propositions. Thus all knowledge representations take place in language. The representational theory studies the mental representation in a formal language called language of thought by Jerry Fodor.

A representation is something that stands for something else. For example, the words of any human language are forms of a representation because they stand for objects, events, and ideas; words are an abstract representation because the relation between a word and the object signified or the idea it represents is arbitrary; words in other language can refer to the same objects and ideas, and in few cases the referent of a word cannot be predicted from its auditory form. But in the case of mental representations, mind preserves information about objects or events in the world. For example, when we have a mental representation of table and this representation preserves locations of objects in the space, it supports a number of abilities, including imaging the place, estimating distance from memory, and so on. This mental representation is necessary because human behaviour cannot be explained without specifying how individuals represent the world to themselves.

According to cognitive scientists, all cognitive processes are computational processes, and all computational processes require internal representations as the medium of computation. And then the nature of this medium is such that the internal representations are symbols. The symbol-system facilitates the computational processes. Computational modelling in cognitive science and artificial intelligence has profoundly affected how human cognition is viewed and studied. Computational levels of cognitions show that the mind is a computer, which is based on symbolic computation. In symbolic computation, the abstractions provided are symbols and rules. But according to the classical computational theory of mind, mental representations are symbolic structures, and the mental processes consist in the manipulation of these representations according to symbolic algorithms as computation is based on symbolic rules.

Herbert Simon, one of the founders of Computer Science, discusses the nature of cognition while constructing models of human mental activity. According to him, cognition is a mental process based on mechanisms of the brain. The brain mechanisms can be studied by neurophysiology. Further, he argues that human brain functions like a computer so that the human cognitive processes are computational in nature. Therefore, if we manage to programme a computer to play chess, we may well have discovered how human thought proceeds. Thus, Simon argues for a sharp distinction between the brain as a physical system and the programs the brain executes; and therefore, he urges us to concentrate our attention on the program. A great deal of the modern study of cognition depends on the insight that representational level and neural level events can be linked through the development of intermediate, computational theories of thought. Newell and Simon point out that this insight is based upon a rather sophisticated notion, both of thinking and computation as activities that are carried out by physical symbol systems.

As we know, a machine or computer is a physical device that manipulates electrical signals that stand for the symbols in the equation. The physical systems translate signal system into a symbol system. Computers and engineer-paper-pencil devices are general computing systems in the sense that they can, in principle, compute any computable function that is defined by a symbol system. In order to actually compute something, the physical device must be given a set of instructions which are stated in terms of symbols. Thus computing systems operate systems of symbols to arrive at results. We have to note that the algorithm is not stated in terms of the physical machine, because the physical operations that achieve the primitive function such as writing down, multiplying, and subtracting have not been specified.

According to Newell, "a physical symbol system has the necessary and sufficient means for general intelligent action." By 'necessary' we mean that any system that exhibits general intelligence will prove upon analysis to be a physical symbol system. By 'sufficient' we mean that any physical symbol system of sufficient size can be organized further to exhibit general intelligence. Lastly, by 'general intelligent action' we wish to indicate the same scope of intelligence as we see in human action. Thus physical symbol systems as discussed above give rise to intelligent actions because of the presence of symbol manipulations according to rules.

The physical symbol system hypothesis plays an important role in showing computational levels of cognition because the symbol system hypothesis implies that the symbolic behaviour of humans arise because he or she has the characteristics of a physical symbol system. Hence, the success in modelling human behaviour on the symbol systems becomes an important part of the evidence for the hypothesis. The hypothesis helps research in cognitive psychology. Research in information processing psychology involves two main kinds of empirical activity. Firstly, it conducts observations and experiments on human behaviour in tasks requiring intelligence. Secondly, it formulates the hypothesis about the symbolic processes found in the human system. Not only are psychological experiments required to test the veridicality of the human behaviour, but also out of experiments come new ideas for the design and construction of physical symbol systems.

According to Newell, Rosenbloom, and Laird, human beings can be described at different levels of the system. For him, at the top is the knowledge level which describes the persons having goals and knowledge about the world, where knowledge is used in the service of attaining goals. The person can operate at the knowledge level because that involves a symbol system, which is a system that operates in terms of representations and information processing operations on this representation. They says, "the symbol level must also be realized in terms of some substrate, that the architecture is that substrate defined in an appropriate descriptive language. For computers this turns out to be the register-transfer level, in which bit-rectors are transported from one functional unit (such as an adder) to another, subject to gating by control bits. For human beings it is neural circuit level, which currently seems well described as highly parallel interconnected networks of inhibitory and exlitatory connections that process a medium of continuous signals."

Thus according to Newell, Rosenbloom, and Laird, the role of architecture in cognitive science is to be the central element in a theory of human cognition. The fixed structure provides the frame within which cognitive processing in the mind takes place. This structure is called architecture. The central function of architecture is to support a system capable of universal computations. Symbols do provide an internal representation function, but representation of the external world is a function of the computational system as a whole so that the architecture supports such representation. It is not the sole or even the predominant determinant of the behaviour of a person, but it is determinant of what makes human behaviour psychological. For him, to have a theory of cognition is to have a theory of the architecture.

The model of the brain, on the other hand, is a technique for analyzing the anatomy and physiology of the brain. This view suggests that the brain consists of a network of simple electrical processing units, which simulate and inhibit one another. This style of explanation of the brain, in cognitive science, is generally considered as the brain-style computation.

Now, the question is: Why should there be a brain-style computation? The basic assumption is that we seek explanation at the program or functional level rather than the implementational level. Thus it is often pointed out that we can learn very little about what kind of program a particular computer may be running by looking at the electronics with which it is made. In fact, we do not care much about the details of the computer at all. All we care about it is the particular program that is running. If we know the program, we will know how the system will behave in any situation. It does not matter whether we use vacuum tubes or transistors, the essential characteristics are the same. It is true for computers because they are all essentially the same. Whether we make them out of vacuum tubes or transistors, we invariably use computers of the same design. But when we look essentially at a difficult architecture, we see that the architecture makes a good deal of difference. It is the architecture that determines what kind of algorithms are most easily carried out on the machine in question. It is the architecture of the machines that determine the essential nature of the program itself. Thus, it is reasonable that we should begin by asking what we know about the architecture of the brain and how it might shape the algorithms underlying the biological intelligence and human mental life.

Rumelhart says that the basic strategy of the connectionist approach is to take the neuron as its fundamental processing unit. Computation is carried out through simple interactions among such processing units. Essentially, the idea is that these processing elements communicate by sending numbers along the lines that connect the processing elements. This identification already provides some interesting constraints on the kinds of algorithms that might underlie the identifications of human intelligence. A question may arise here: How does the replacement of the computer metaphor as model of mind affect our thinking? Rumelhart says that this change in orientation leads us to a number of considerations that further inform and constrain our model building efforts. Neurons are remarkably relative to the components in modern computers. These neurons operate in the time scale of milliseconds, whereas computer components operate in the time scale of nanoseconds- a vector of 106 time faster. This means that the human brain process that receives the order in a second or less can involve only a hundred or so times steps. Because, most of the computational processes like perception, memory retrieval, etc take about a second to function. That is, we seek explanations for these mental phenomena that do not require more than about a hundredth elementary sequential operations.

The human brain contains billions of such processing elements. As the computer organizes computation with many serial steps, similarly the brain can deploy many processing elements in cooperation and in parallel to carry out its activities. Thus, the use of brain style computational system offers not only a hope that we can characterize how brains actually carry out certain information processing tasks but also offers solution to computational problems that seem difficult to solve in more traditional computational framework.

The connectionist systems are capable of exploiting and mimicking brain-style computation like artificial intelligence. Connectionism operates both as a system and a process. The connectionist systems are very important because they provide good solutions to a number of difficult computational problems that seem to arise often in models of cognition. Connectionism as a processing mechanism is carried out by a number of processing elements. These elements, called nodes or units, have a dynamics, which is roughly an analogue to simple neurons. Each node receives input from some number of the nodes and responds to that input according to a simple activation function, and in turn excites or inhibits other nodes to which it is connected.

All the levels of cognition, which we have discussed, are not universally accepted. There are levels of cognition which are better explained in the connectionist model. The connectionist approach refers to the functions of the neurons. According to Crick, the 'astonishing hypothesis' is that consciousness and all the thinking that goes with it are the product of neurons, and therefore, the primary business of the scientific study of thought has to be the reduction of cognitive psychology to neurology. He says, "One may conclude, then, that to understand the various forms of consciousness, we need to know their neural correlates."

Crick is not opposed to the study of the computational level of cognition when they are linked to the physiological investigations of the brain. But, advocates of computational level models are careful to dissociate themselves from the idea that the computer, as a physical device, is being used as a model for the brain.

According to David Chalmers, a computational basis for cognition can be challenged in two ways. First, it can be argued that computation cannot do what cognition does: that a computational simulation cannot reproduce a human cognition, because the causal structure in human cognition goes beyond what a computational description can do. Second, computation might capture the human capacities, but something more is required to replace the human capacity.

The human cognition can be applied to what is known as memory attention, pattern recognition, language, problem solving etc. The most intriguing aspect of the human mind is the selection of information for further processing and storage. The information available during each moment, except sleep and unusual occasions, is vast and complex; we are constantly bombarded by our senses. All the external senses give us information which the mind deals with at different stages. Cognition is the output produced after a long process of getting the inputs.

## Folk Psychology

There are different models of mind: connectionism is one of them. In the modern cognitive science, these models have provided the basis for simulating or modeling cognitive performance. Simulation is one of the important ways of testing theories of the mind. If a simulation performs in a manner comparable to the mind, then that will offer support for the theory underlying that simulation.

However, in cognitive science, two models have provided the basis for most of the simulation activity. On the one hand, the digital computer can be used to manipulate symbols. In so far as it becomes possible to program the symbol-processing computer to execute tasks that seems to require intelligence, the symbol-processing computer becomes a plausible analogy to the mind. Numerous cognitive science theorists have been attracted to the proposal that the mind itself is a symbol-processing device.

The model of the brain, on the other hand, is a technique for analyzing the anatomy and physiology of the brain. This view suggests that the brain consists of a network of simple electrical processing units, which simulate and inhibit one another. This style of explanation of the brain, in the cognitive science, is generally considered as the brain-style computation.

Now, the question is: why should there be a brain-style computation? The basic assumption is that we seek explanation at the program or functional level rather than the implementational level. Thus it is often pointed out that we can learn very little about what kind of program a particular computer may be running by looking at the electronics with which it is made. In fact, we do not care much about the details of the computer at all: all we care about is the particular program that is running. If we know the program, we will know how the system will behave in any situation. It does not matter whether we use vacuum tubes or transistors: the essential characteristics are the same. It is true for computers because they are all essentially the same. Whether we make them out of vacuum tubes or transistors, we invariably use computers of the same design. But when we look at essentially a difficult architecture, we see that the architecture makes a good deal of difference. It is the architecture that determines which kinds of algorithms are most easily carried out on the machine in question. It is the architecture of the machines that determines the essential nature of the program itself. Thus, it is reasonable that we should begin by asking what we know about the architecture of the brain and how it might shape the algorithms underlying the biological intelligence and human mental life.

Rumelhart says that the basic strategy of the connectionist approach is to take the neuron as its fundamental processing unit. We imagine that computation is carried out through simple interactions among such processing units. Essentially, the idea is that these processing elements communicate by sending numbers along the lines that connect the processing elements. This identification already provides some interesting constraints on the kinds of algorithms that might underlie human intelligence. A question may arise here: How does the replacement of the computer metaphor as model of mind affect our thinking? Rumelhart says that this change in orientation leads us to a number of considerations that further inform and constrain our model building effort.

Because, neurons are remarkably relative to the components in modern computers. Neurons operate in the time scale of milliseconds, whereas computer components operate in the time scale of nanoseconds- a vector of 106 time faster[2]. This means that human brain process that receives the order in a second or less can involve only a hundred or so times steps. Because, most of the computational processes like perception, memory retrieval, etc take about a second to function. That is, we seek explanations for these mental phenomena that do not require more than about a hundredth elementary sequential operations.

The human brain contains billions of such processing elements. As the computer organizes computation with many serial steps, similarly the brain can deploy many processing elements cooperatively and in parallel to carry out its activities. Thus, the use of brain style computational system offers not only a hope that we can characterize how brains actually carry out certain information processing tasks but also offers solution to computational problems that seem difficult to solve in more traditional computational framework.

The connectionist systems are capable of exploiting and mimicking brain-style computation like artificial intelligence. Connectionism operates both as a system and a process. The connectionist systems are very important because they provide good solutions to a number of difficult computational problems that seem to arise often in models of cognition. Connectionist model can solve best-mach-search, rapid-pattern-matching, implementing content-addressable memory-storage systems and this model allow many more to its environment. Connectionism as a processing mechanism is carried out by a number of processing elements. These elements, called nodes or units, have a dynamics, which is roughly an analogue to simple neurons. Each node receives input from some number of the nodes and responds to that input according to a simple activation function, and in turn excites or inhibits other nodes to which it is connected.

## The Connectionist Framework

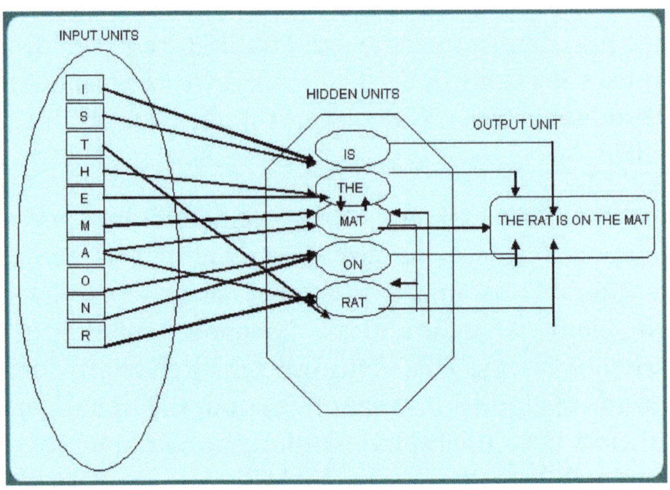


Connectionist model

## The Connectionist Framework

The above figure is arbitrarily taken as a connectionist model. In any connectionist model, there are three units such as input units, hidden units, and output units. Here, the input units are such as 'I', 'S', .... 'N', 'R', the hidden units are 'IS', 'THE', 'MAT', 'ON', 'RAT', and the output unit is "THE RAT IS ON THE MAT." There may be many inputs, hidden units, and many output units. The hidden units serve neither as input nor output units, but facilitate the processing of information through the system. This model will be very clear, if we go through Rumelhart's seven major components of any connectionist model.

i. A set of processing units

Any connectionist system begins with a set of processing units. All of the processing of connectionist system is carried out by these units. There is no executive or other agency. There are only relatively simple units, each doing its own relatively simple job. A unit's job is simply to receive input from its neighours and, as its function, it sends output values to its neighbors. The system is inherently parallel in the sense that many units can carry out their computations at the same time.

There are three types of units, input, output, and hidden units. Input units receive input from sources external to the system under study. The output units send signal out of the system. The hidden units are those that check that inputs and outputs are from within the system they are modeling. They are not visible to outside system.

ii. The state of activation

In addition to the set of units we also need a representation of the state of the system at time 'T'. This is primarily specified by a vector (T), representing the pattern of activation over the set of processing units. Each element of the vector stands for the elements of one of the units. It is the pattern of activation over the set of units that captures what the system represents at any time. It is useful to see processing in the system as the evolution, through time, of a pattern of activity over the set of units.

iii. Output of the units

Units interact by transmitting signals to their neighbors. The strength of their signals and the degree to which they affect their neighbors are determined by their degree of activation. But in some of our models the output level is exactly equal to the activation levels of the unit. The output of the unit depends on its activation values.

iv. The pattern of connectivity

Units are connected to one another. It is this pattern of connectivity that constitutes what the system knows and determines how it will respond to any arbitrary input. Specifying the processing system and the knowledge encoded therein is a matter of specifying this pattern of connectivity among the processing units.

v. Activation rules

We also need to set of rules whereby the inputs impinging on a particular unit are combined with one another and undergoing processing with the current states of unit produces a new state of activation.

vi. Modifying pattern of connectivity as a foundation of experience

Changing the processing or knowledge structure in a connectionist system involves modifying the pattern of interconnectivity. Generally, there are three kinds of modifications:

i. Development of new connections

ii. Loss of existing connections.

iii. Modification of the strength of connection that already exists.

vii. Representation of the environment

For the development of any model, it is very difficult to have a clear representation of the environment in which this model is to exist. In connectionist model, we represent the environment as a time-making stochastic function over the space of input patterns. That is, we imagine that at any point of time there is some probability that any of the possible set of input patterns is impinging on the input units. This probability depends on the history of inputs as well as outputs of the system. In practice, most models involve a much simpler characterization of the environment.

## Connectionism and Its Relation with Other Discipline

Now the question is: What is the relation of connectionism with other disciplines like artificial intelligence, and philosophy of mind? Cognitive science is an interdisplinary research area, which emerged from the cognitive revolution. Cognitive science includes artificial intelligence, cognitive psychology, linguistics, neuroscience and philosophy. It reveals functional unity among diverse epistemological assumptions because they share certain core assumptions of the symbolic approach to cognition. But where as connectionism is related to all these areas because connectionism is a part of neuroscience, which talks about cognition in a different manner. This is said to be done by locating the neurons in the cerebral cortex that correlates best with consciousness and figuring out how they link to neurons elsewhere in the brain as the connectionist explains in the same way. This theory is first outlined by Crick and Koch. They hypothesize that these oscillation are the basis of consciousness. This is partly because the oscillations seem to be correlated with awareness in different modalities within the visual and olfactory systems and also because they suggest a mechanism by which the binding of information contents might be achieved. Binding is the process whereby separately represented pieces of information about a single entity are brought together to be used by later processing, as when information about the colour and shape of a perceived objected is integrated from separate visual pathways. Both connectionists and neuroscientists are exploring consciousness or mind broadly from a materialistic point of view. They leave out the essence of mind, and forget about the really difficult aspects. Now, we may raise some of the questions like why does it exist? What does it do? How could it possibly arise from pulpy gray matter? How can an unintelligent machine could give rises an intelligent experience? If the cognitive scientists try to give an answer, then that answer will not be an appropriate for the relevant questions. But it is very difficult to give precise definitions of mind.

## Connectionism and Artificial Intelligence

Artificial intelligence has witnessed the emergence of several new methods of analysis. These include connectionism that investigates the properties of networks of neurons like units. This approach focuses on computational methods inspired by natural phenomena. Connectionism is inspired by observations of basic neural activity in biological organisms. Connectionism is an approach to cognitive modeling, which, in contemporary usage, refers to particular classes of computer-implemented models of artificial intelligence. Artificial intelligence gives importance to the mind, where as connectionism puts emphasis on the brain. For connectionism, human brain is a neural network; that is to say that there is a relation among the neurons. Artificial intelligence argues that the mind is the software, and the brain is the hardware in which the mind works and this is the view of functionalism. Thus both connectionism and artificial intelligence belong to the same theory concerned about human mind.

## Philosophical Implication of Connectionism

In the understanding of cognition, connectionism will necessarily have implication of philosophy of mind. There are two areas in particular on which it is likely to have impact. They are the analysis of the mind as a representational system and the analysis of intentational representational.

Fodor distinguishes the computational theory of mind from the representational theory of mind. The representational theory of mind holds the view that systems have mental states by virtue of encoding representations and standing in particular relations to them. The computational theory adds that cognitive activity consists of formal operations performed on these representations.

Fodor and Pylyshyn's argument against connectionism brings out the defect of the connectionist model. They opine that it fails to support the computational theory. Fodor interprets connectionist models as representational and so potentially conforming to the representational theory of mind. This is because connectionists routinely interpret the activations of units or groups of units as representing contents. This is the case for input and output units providing cognitive interpretation of a network's activity, thus a theorist must treat the input as a representation of a problem and the output as representing the answer. Sometimes this is done unit by unit. A given unit is found to be activated by inputs with certain features and so interpreted as representing those features. This is interpreted as that the network has differentiated inputs with differentiated features. This suggests that connectionist systems can indeed be understood as the representation theory of mind. Even if connectionist networks exemplify the representational theory of mind, they are significantly different from more traditional examples of the representational theory.

Firstly, it is not clear that we can always give an interpretation to what units in a connectionist network represent in natural language terms.

Secondly, the representations that are constructed are not discrete but distributed. That is, the same units and the same connections connect many different representational roles rather than employing one representation per role. This distinguishes connectionists' representations from those that have previously been designed.

Thirdly, it is emphasized that the pattern of activations on hidden units in connectionist systems are the products of the learning that the system has undergone. The interpretations assigned to these units are not arbitrary. They are represented symbolically, but are analysis of how the network has solved the problem it was confronting. Thus the network is connected to real sensory inputs, and not supplied inputs by the modeler machinery. The intentationality of these representations is genuine, not merely a product of the theorist's interpretation.

In philosophy of mind and cognitive science, folk psychology, or commonsense psychology, is a human capacity to explain and predict the behavior and mental state of other people. Processes and items encountered in daily life such as pain, pleasure, excitement, and anxiety use common linguistic terms as opposed to technical or scientific jargon.

Traditionally, the study of folk psychology has focused on how everyday people—those without formal training in the various academic fields of science—go about attributing mental states. This domain has primarily been centred on intentional states reflective of an individual's beliefs and desires; each described in terms of everyday language and concepts such as "beliefs", "desires", "fear", and "hope".

Eliminative materialism is the claim that folk psychology is false and should be discarded (or "eliminated").

## Key Folk-concepts

### Intentionality

When perceiving, explaining, or criticizing human behaviour, people distinguish between intentional and unintentional actions. An evaluation of an action as stemming from purposeful action or accidental circumstances is one of the key determinants in social interaction. Others are the environmental conditions or pre-cognitive matters. For example, a critical remark that is judged to be intentional on the part of the receiver of the message can be viewed as a hurtful insult. Conversely, if considered unintentional, the same remark may be dismissed and forgiven.

This folk concept of intentionality is used to distinguish between intentional and unintentional behaviour in sports where intentional fouls are punished more harshly than ones deemed to be unintentional. It is also applied in the legal system in terms of criminal law distinguishing between murder and manslaughter.

The importance of this concept transcends almost all aspects of everyday life: with empirical studies in social and developmental psychology exploring perceived intentionality's role as a mediator for aggression, relationship conflict, judgments of responsibility blame or punishment.

Recent empirical literature on folk psychology has shown that people's theories regarding intentional actions involve four distinct factors: beliefs, desires, causal histories, and enabling factors. Here, beliefs and desires represent the central variables responsible for the folk theories of intention.

Desires embody outcomes that an individual seeks, including those that are impossible to achieve. The key difference between desires and intentions is that desires can be purely hypothetical, whereas intentions specify an outcome that the individual is actually trying to bring to fruition.

In terms of beliefs, there are several types that are relevant to intentions—outcome beliefs and ability beliefs. Outcome beliefs are beliefs as to whether a given action will fulfill an intention, as in "purchasing a new watch will impress my friends". Ability consists of an actor's conviction regarding his or her ability to perform an action, as in " I really can afford the new watch". In light of this, Heider postulated that ability beliefs could be attributed with causing individuals to form goals that would not otherwise have been entertained.

## Comprehension and Prediction

### The Context Model

Folk psychology is crucial to evaluating and ultimately understanding novel concepts and items. Developed by Medin, Altom, and Murphy, the Context Model hypothesizes that as a result of mental models in the form of prototype and exemplar representations, individuals are able to more accurately represent and comprehend the environment around them.

According to the model, the overall similarity between the prototype and a given instance of a category is evaluated based on multiple dimensions (e.g., shape, size, color). A multiplicative function modeled after this phenomenon was created.

$$s(P,E_i) = \prod_k s(P,E_{ik})$$

Here, *S(P,Ei)* represents the similarity between the prototype and the *i*th exemplar, *k* is the subscript for the dimensions *(k = 1...k)*, and *S(P,Eik)* is the similarity between the prototype and the *i*th exemplar on the *k*th dimension.

## Explanation

## Hilton's (1990) Conversational Model

Given that folk psychology represents causal knowledge associated with the mind's categorization processes, it follows that folk psychology is actively employed in aiding the explanation of everyday actions. Denis Hilton's (1990) Conversational Model was created with this causal explanation in mind, with the model having the ability to generate specific predictions. Hilton coined his model the 'conversational' model because he argued that as a social activity, unlike prediction, explanation requires an audience: to whom the individual explains the event or action. According to the model, causal explanations follow two particular conversational maxims from Grice's (1975) models of conversation—the manner maxim and the quantity maxim. Grice indicated that the content of a conversation should be relevant, informative, and fitting of the audience's gap in knowledge. Cognizant of this, the Conversational Model indicates that the explainer, upon evaluation of his audience, will modify his explanation to cater their needs. In essence, demonstrating the inherent need for mental comparison and in subsequent modification of behaviour in everyday explanations.

## Application and Functioning

## Belief-desire Model

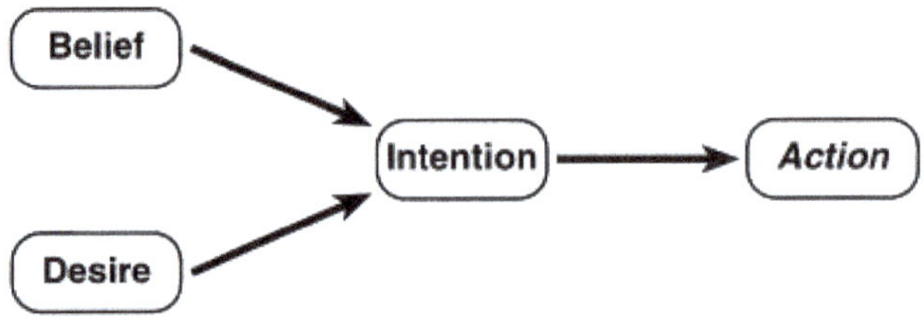


A schematic representation of folk psychology of belief, desire, intention, and action

The belief-desire model of psychology illustrates one method in which Folk Psychology is utilized in everyday life. According to this model, people perform an action if they want an outcome and believe that it can be obtained by performing the action. However, beliefs and desires are not responsible for immediate action; intention acts as a mediator of in this relationship belief-desire and action. In other words, a person who wants to achieve a goal, "G", and believes action "A" will aid in attaining "G"; this leads to an intention to perform "A", which is then carried out to produce action "A".

Schank & Abelson (1977) described this inclusion of typical beliefs, desires, and intentions underlying an action as akin to a "script" whereby an individual is merely following an unconscious framework that leads to the ultimate decision of whether an action will be performed. Similarly, Barsalou (1985) described the category of the mind as an "ideal" whereby if a desire, a belief, and an intention were all present, they would "rationally" lead to a given action. They coined this phenomenon the Ideal of Rational Action.

## Goal-intentional Action Model

Existing literature has widely corroborated the fact that social behavior is greatly affected by the causes to which people attribute actions. In particular, it has been shown that an individual's interpretation of the causes of behaviour reflects their pre-existing beliefs regarding the actor's mental state and motivation behind his or her actions. It follows that they draw on the assumed intentions of actors to guide their own responses to punish or reward the actor. This concept is extended to cover instances in which behavioural evidence is lacking. Under these circumstances, it has been shown that the individual will again draw on assumed intentions in order to predict the actions of the third party.

Although the two components are often used interchangeably in common parlance, there is an important distinction between the goals and intentions. This discrepancy lies in the fact that individuals with an intention to perform an action also foster the belief that it will be achieved, whereas the same person with a goal may not necessarily believe that the action is able to be performed in spite of having a strong desire to do so.

Predicting goals and actions, much like the Belief-Desire Model, involves moderating variables that determine whether an action will be performed. In the Goal-Intentional Action Model, the predictors of goals and actions are: the actors' beliefs about his or her abilities and their actual possession of preconditions required to actually carry out the action. Additionally, preconditions consist of the various conditions necessary in order for realization of intentions. This includes abilities and skills in addition to environmental variables that may come into play. Schank & Abelson raises the example of going to a restaurant, where the preconditions include the ability to afford the bill and get to the correct venue, in addition to the fact that the restaurant must be open for business. Traditionally, people prefer to allude to preconditions to explain actions that have a high probability of being unattainable, whereas goals tend to be described as a wide range of common actions.

## Model of Everyday Inferences

Models of everyday inferences capture folk psychology of human informal reasoning. Many models of this nature have been developed. They express and refine our folk psychological ways of understanding of how one makes inferences.

For example, one model describes human everyday reasoning as combinations of simple, direct rules and similarity-based processes. From the interaction of these simple mechanisms, seemingly complex patterns of reasoning emerge. The model has been used to account for a variety of reasoning data.

## Controversy

Folk psychology remains the subject of much contention in academic circles with respect to its scope, method and the significance of its contributions to the scientific community. A large part of this criticisms stems from the prevailing impression that folk psychology is a primitive practice reserved for the uneducated and non-academics in discussing their everyday lives.

There is significant debate over whether folk psychology is useful for academic purposes; specifically, whether it can be relevant with regard to the scientific psychology domain. It has been argued that a mechanism used for laypeople's understanding, predicting, and explaining each other's actions is inapplicable with regards to the requirements of the scientific method. Conversely, opponents have called for patience, seeing the mechanism employed by laypeople for understanding each other's actions as important in their formation of bases for future action when encountering similar situations. Malle & Knobe hailed this systematization of people's everyday understanding of the mind as an inevitable progression towards a more comprehensive field of psychology. Medin et al. provide another advantage of conceptualizing folk psychology with their Mixture Model of Categorization: it is advantageous due to its facilitation of action prediction.

## Whether Connectionism Contributes to the Replacement of Folk-psychology

We know that in many ways cognitive science originated from philosophy. The importance of connectionism to philosophy emerges first with respect to the question of whether folk-psychology remains viable or must be replaced. If it is replaced, then the reliance on prepositional representations of knowledge in other areas of philosophy would be at risk. Because connectionism explains mind in terms of mechanical processes, it leaves out the 'mentality' of the human mind. This theory suggests that there is no mental quality such as belief, intention, etc. If connectionism should provide a correct account of mental processing, and if it did not turn out to implement symbolic systems, then the account of mental life as actually involving the manipulation of propositions would appear to be false. That is, mental states involving propositions will not figure in the causal genesis of behavior.

William Bechtel and Adele Abrahamsen quote from William Churchland that eliminative materialism by maintaining that if a theory fails to reduce to our best scientific theories at lower levels, it must be dismissed as false. They contend that reduction fails in the case of folk psychology because there is nothing in the head with which to identify the propositions it posits, this conclusion entails the further conclusion that folk psychology is false. In making this inference, they assume that folk-psychological theories about processes occur inside people's mind. Now, we have to examine the question whether connectionism contributes to the replacement of folk-psychology.

According to William Bechtel and Adele Abrahamsen, folk psychology refers to people's attributions of prepositional attributes to other people and uses these to predict and explain their behaviour. These attributions are made to whole persons; folk psychology does not itself offer an account of the finer-gained internal operations that may produce prepositional attitudes. If we attribute to a person a particular belief that itself need not be a discrete internal state; the states inside the person that enable the person to have a belief will have a quite different character.

They apply the above point to the case of cognition. The activities inside the head may make it possible for a person to have beliefs and desires, but it does not assume that they have internal states that correspond to these prepositional attributes. It may be that the internal activities are best described in the connectionist approach. However, it does not show that folk psychology is false. But if it is false, it will be so because it does not give a correct characterization of the cognitive state of persons and must, therefore, be replaced by a better theory at the same level.

Here, I would like to argue that the connectionists' model of mind is unable to refute folk-psychology. The connectionists explain mind in terms of syntactic, and thereby neglect semantics, which is very important for understanding the human mind. There is mental content which represent the world, that is to say that there is central agency or the 'I' to which the mental activity is ascribed. This shows that human mind has propositional attitudes about the world. As David Chalmers pointed out which are the mental states such as 'belief', 'doubt', etc., often called 'propositional attitudes' are attitudes to propositions concerning the world. For example, when I believe that John will tour India, I endorse a certain proposition concerning John; when I hope that John will tour India, I have different attitudes toward the same proposition. Here, the central feature of these mental states is their semantic aspect, or intentationality. That is, a belief has semantic content, the content of my belief cited is something like the proposition that John will tour India. This semantic or intentationality aspect has the features of subjectivity and qualia. The subjectivity of consciousness is an essential feature of mental states, which can prove that the ontology of mental states is an irreducible fact of first person ontology. Where as in the case of connectionists model of mind, there is no subjective experience, and it gives the explanations of mind in the third-person perspective.

Now the question is: Can the subjective experience be made a part of the objective structure of the natural order in the way the connectionist functions of the mind are? This has generated a debate as to whether there can be a complete reduction of the subjective experience into mechanical states of the brain. William Bechetel, Rumelhart, and Marr are fully committed to the replacement of folk-psychology. However, this can be opposed on the ground that the mental beliefs are ascribed to a conscious subject and not to the connectionist model of mind or brain because the brain is at best a physical system though with infinite physical capacity. The subject is non-reducible to the brain in the sense that the brain itself belongs to the subject. The subject functions autonomously; the qualia as well as the brain states are only different states of the autonomous subject. Thus the reality of the subject of the qualia has to be admitted if we can have coherent theory of mind.

The connectionist model of mind fails to account for the real nature of the mental states because of its not too clear attempt to reduce mental states to the machine-states. Connectionist fails as a theory of mind because of its reductionist dogma: It makes mind superfluous in the universe.15 Mind is made at best a mechanical system with certain determinate functions.

The mechanistic theory of mind in all hues faces the question as to how we can account for the qualitative content of our consciousness. It cannot ultimately tell us how the subjective experience is possible and how consciousness can be real in the universe. The mechanistic view does not have any convincing answer to the question how qualia are a necessary feature of consciousness. If mind functions like a machine, it can at best exhibit only mechanical states which look very much like the mental states but ontologically are very different from them.

## Representations

Representations are about things other than themselves and are intentional in the sense of being about 'this' or 'that'. Because the mental representations have content, which is related to belief, intention, thought, and action, they are also intentional in the sense of being purposive. Now we may ask: What is it that distinguishes items that serve as representations from other objects or events? And what distinguishes the various kinds of symbols from one another? As for the first question, there has been general agreement that the basic notion of a representation involves things like 'standing for', 'being about', 'referring to', and 'undenoting' something else. Some theorists have maintained that it is only the use of symbols that exhibits or indicates the presence of mind and mental states.

Mental representation, like beliefs and thoughts, constitutes the broad domain of cognitive science. They explain how cognition takes place in the human mind. Cognitive science (including cognitive linguistics and cognitive psychology) has brought about a cognitive revolution in the study of mind. Here, we can undertake two important developments in cognitive science. One is the representational theory of mind. For, to accept the representational theory of mind is to accept that mental representations are very much like the interrepresentational states of a digital computer. The other is the adoption of a computational model of mind or computational theory of mind.

In turn, two questions have to be answered in this connection: What kinds of representational systems are employed in cognition? What is machine intelligence or artificial intelligent? Fodor has answered these questions in his computational representational theory of mind (CRTM in short). The computational representational theory of mind makes a strong assumption about mental processes: Mental processes are computational processes, i.e., formal operations defined over symbols. In Fodor's view, "computational processes are both symbolic and formal. They are symbolic because they are defined over representations, and they are formal because they apply to representations in virtue of (roughly) the syntax of the representations." The theory purports to offer a solution to the problem raised by the compositionality of propositional attitudes like beliefs, thoughts, etc.; secondly, it proposes to vindicate the strong reading of the intentional realist casual thesis regarding the mental phenomena. Again, it may be noted that the CRTM is consequently based on two fundamental assumptions; the first is Fodors' Language of Thought (LoT) hypothesis, and the second is the view that psychological explanation that is both intentional and nomological.

## Mental Representation

A mental representation (or cognitive representation), in philosophy of mind, cognitive psychology, neuroscience, and cognitive science, is a hypothetical internal cognitive symbol that represents external reality, or else a mental process that makes use of such a symbol: "a formal system for making explicit certain entities or types of information, together with a specification of how the system does this."

Mental representation is the mental imagery of things that are not actually present to the senses. In contemporary philosophy, specifically in fields of metaphysics such as philosophy of mind and ontology, a mental representation is one of the prevailing ways of explaining and describing the nature of ideas and concepts.

Mental representations (or mental imagery) enable representing things that have never been experienced as well as things that do not exist. Think of yourself traveling to a place you have never visited before, or having a third arm. These things have either never happened or are impossible and do not exist, yet our brain and mental imagery allows us to imagine them. Although visual imagery is more likely to be recalled, mental imagery may involve representations in any of the sensory modalities, such as hearing, smell, or taste. Stephen Kosslyn proposes that images are used to help solve certain types of problems. We are able to visualize the objects in question and mentally represent the images to solve it.

Mental representations also allow you to experience things right in front of you—though the process of how the brain interprets the representational content is debated.

## Representational Theories of Mind

Representationalism (also known as indirect realism) is the view that representations are the main way we access external reality. Another major prevailing philosophical theory posits that concepts are entirely abstract objects.

The representational theory of mind attempts to explain the nature of ideas, concepts and other mental content in contemporary philosophy of mind, cognitive science and experimental psychology. In contrast to theories of naive or direct realism, the representational theory of mind postulates the actual existence of mental representations which act as intermediaries between the observing subject and the objects, processes or other entities observed in the external world. These intermediaries stand for or represent to the mind the objects of that world.

For example, when someone arrives at the belief that his or her floor needs sweeping, the representational theory of mind states that he or she forms a mental representation that represents the floor and its state of cleanliness.

The original or "classical" representational theory probably can be traced back to Thomas Hobbes and was a dominant theme in classical empiricism in general. According to this version of the theory, the mental representations were images (often called "ideas") of the objects or states of affairs represented. For modern adherents, such as Jerry Fodor, Steven Pinker and many others, the representational system consists rather of an internal language of thought (i.e., mentalese). The contents of thoughts are represented in symbolic structures (the formulas of Mentalese) which, analogously to natural languages but on a much more abstract level, possess a syntax and semantics very much like those of natural languages. For Augusto (2014), at this abstract, formal level, the syntax of thought is the set of symbol rules (i.e., operations, processes, etc. on and with symbol structures) and the semantics of thought is the set of symbol structures (concepts and propositions). Content (i.e., thought) emerges from the meaningful co-occurrence of both sets of symbols that, in turn, is determined by the semantic protoness of syntax and the syntactic protoness of semantics. For instance, "8 x 9" is a meaningful co-occurrence, whereas "CAT x" is not; "x" is a symbol rule called for by symbol structures such as "8" and "9", but not by "CAT".

## Strong vs weak, Restricted vs Unrestricted

There are two types of representationalism, strong and weak. Strong representationalism attempts to reduce phenomenal character to intentional content. On the other hand, weak representationalism claims only that phenomenal character supervenes on intentional content. Strong representationalism aims to provide a theory about the nature of phenomenal character, and offers a solution to the hard problem of consciousness. In contrast to this, weak representationalism does not aim to provide a theory of consciousness, nor does it offer a solution to the hard problem of consciousness.

Strong representationalism can be further broken down into restricted and unrestricted versions. The restricted version deals only with certain kinds of phenomenal states e.g. visual perception. Most representationalists endorse an unrestricted version of representationalism. According to the unrestricted version, for any state with phenomenal character that state's phenomenal character reduces to its intentional content. Only this unrestricted version of representationalism is able to provide a general theory about the nature of phenomenal character, as well as offer a potential solution to the hard problem of consciousness. The successful reduction of the phenomenal character of a state to its intentional content would provide a solution to the hard problem of consciousness once a physicalist account of intentionality is worked out.

## Problems for the Unrestricted Version

When arguing against the unrestricted version of representationalism people will often bring up phenomenal mental states that appear to lack intentional content. The unrestricted version seeks to account for all phenomenal states. Thus, for it to be true, all states with phenomenal character must have intentional content to which that character is reduced. Phenomenal states without intentional content therefore serve as a counterexample to the unrestricted version. If the state has no intentional content its phenomenal character will not be reducible to that state's intentional content, for it has none to begin with.

A common example of this kind of state are moods. Moods are states with phenomenal character that are generally thought to not be directed at anything in particular. Moods are thought to lack directedness, unlike emotions, which are typically thought to be directed at particular things e.g. you are mad *at* your sibling, you are afraid *of* a dangerous animal. People conclude that because moods are undirected they are also nonintentional i.e. they lack intentionality or aboutness. Because they are not directed at anything they are not about anything. Because they lack intentionality they will lack any intentional content. Lacking intentional content their phenomenal character will not be reducible to intentional content, refuting the representational doctrine.

Though emotions are typically considered as having directedness and intentionality this idea has also been called into question. One might point to emotions a person all of a sudden experiences that do not appear to be directed at or about anything in particular. Emotions elicited by listening to music are another potential example of undirected, nonintentional emotions. Emotions aroused in this way do not seem to necessarily be about anything, including the music that arouses them.

## Responses

In response to this objection a proponent of representationalism might reject the undirected nonintentionality of moods, and attempt to identify some intentional content they might plausibly be thought to possess. The proponent of representationalism might also reject the narrow conception of intentionality as being directed at a particular thing, arguing instead for a broader kind of intentionality.

There are three alternative kinds of directedness/intentionality one might posit for moods.

- Outward Directedness: What it is like to be in mood M is to have a certain kind of outwardly focused representational content.
- Inward Directedness: What it is like to be in mood M is to have a certain kind of inwardly focused representational content.
- Hybrid Directedness: What it is like to be in mood M is to have both a certain kind of outwardly focused representational content and a certain kind of inwardly focused representational content.

In the case of outward directedness moods might be directed at either the world as a whole, a changing series of objects in the world, or unbound emotion properties projected by people onto things in the world. In the case of inward directedness moods are directed at the overall state of a person's body. In the case of hybrid directedness moods are directed at some combination of inward and outward things.

## Further Objections

Even if one can identify some possible intentional content for moods we might still question whether that content is able to sufficiently capture the phenomenal character of the mood states they are a part of. Amy Kind contends that in the case of all the previously mentioned kinds of directedness (outward, inward, and hybrid) the intentional content supplied to the mood state is not capable of sufficiently capturing the phenomenal aspects of the mood states. In the case of inward directedness, the phenomenology of the mood does not seem tied to the state of one's body, and even if one's mood is reflected by the overall state of one's body that person will not necessarily be aware of it, demonstrating the insufficiency of the intentional content to adequately capture the phenomenal aspects of the mood. In the case of outward directedness, the phenomenology of the mood and its intentional content do not seem to share the corresponding relation they should given that the phenomenal character is supposed to reduce to the intentional content. Hybrid directedness, if it can even get off the ground, faces the same objection.

## Philosophers

There is a wide debate on what kinds of representations exist. There are several philosophers who bring about different aspects of the debate. Such philosophers include Alex

Morgan, Gualtiero Piccinini, and Uriah Kriegel—though this is not an exhaustive list.

## Alex Morgan

There are "job description" representations. That is representations that (1) represent something—have intentionality, (2) have a special relation—the represented object does not need to exist, and (3) content plays a causal role in what gets represented: e.g. saying "hello" to a friend,giving a glare to an enemy.

Structural representations are also important. These types of representations are basically mental maps that we have in our minds that correspond exactly to those objects in the world (the intentional content). According to Morgan, structural representations are not the same as mental representations—there is nothing mental about them: plants can have structural representations.

There are also internal representations. These types of representations include those that involve future decisions, episodic memories, or any type of projection into the future.

## Gualtiero Piccinini

In Gualtiero Piccinini's forthcoming work, he discusses topics on natural and nonnatural mental representations. He relies on the natural definition of mental representations given by Grice (1957) where *P entails that P.* e.g. Those spots mean measles, entails that the patient has measles. Then there are nonnatural representations: *P does not entail P.* e.g. The 3 rings on the bell of a bus mean the bus is full—the rings on the bell are independent of the fullness of the bus—we could have assigned something else (just as arbitrary) to signify that the bus is full.

## Uriah Kriegel

There are also objective and subjective mental representations. Objective representations are closest to tracking theories—where the brain simply tracks what is in the environment. If there is a blue bird outside my window, the objective representation is that of the blue bird. Subjective representations can vary person-to-person. For example, if I am colorblind, that blue bird outside my window will not *appear* blue to me since I cannot represent the blueness of blue. The relationship between these two types of representation can vary.

a. Objective varies, but the subjective does not: e.g. brain-in-a-vat

b. Subjective varies, but the objective does not: e.g. color-inverted world

c. All representations found in objective and none in the subjective: e.g. thermometer

d. All representations found in subjective and none in the objective: e.g. an agent that experiences in a void.

Eliminativists think that subjective representations don't exist. Reductivists think subjective representations are reducible to objective. Non-reductivists think that subjective representations are real and distinct.

The debates about mental representation do not end here. In fact, they lead to other discussions such as content of mental states, embodied cognition, consciousness, and phenomenology.

## Representational Theory of Mind (RTM)

The representational theory of mind shows that any propositional attitude, such as belief, desire, doubt, etc, is literally a computational relation between an organism and some formula in the internal code of that organism. Fodor's view is, as Lycon states it, "to believe that such and such is to have a mental symbol that means that such and such is tokened in your head in a certain way; it's to have such a token 'in your belief box', as I shall sometimes say." It is in virtue of this system for representing and processing information that mental states are related causally to one another.

Moreover, according to Fodor, "propositional attitudes as relations between organisms and internal representations; precisely the view that the psychologists have independently arrived at." By providing a relational treatment of the propositional attitudes, it is possible to state how they are contentful. For the relational treatment, propositional attitudes are dyadic relations and are the internal mental representations. The belief ascription statements are true, only if the organism stands in believed relation to the representational contents. First, it is naturally believable that propositional attitudes are relations. For example, when John believes in something, it seems that John stands in relation to something, that is, the object of the belief. Secondly, existential generalization applies to the verbs of propositional attitudes. For example, if John believes 'it is raining', then we can undoubtedly say that there is something that John believes, which shows that the belief is a relation between John and something that he believes.

Fodor is realistic about intentionality and propositional attitudes. The main point of his theory of intentionality is that intentionality is primary and is originally a real feature in the brain. Language is intentional only in a secondary sense, for it is intentional only in so far as some of the sentences, which are uttered in our natural language, and describe the real features of the mind of oneself or others. It is this fact which helps us in generating our natural languages, and also the special vocabulary in language that involves employment of verbs operating over a propositional content. For example, if we said that John decided to stay at the bus stop, rather than make a run to the local police station, because he believed certain things had certain desires, came up with a certain evaluative decisions, then we are describing a series of real processes in John's brain which involve computational operations over propositional contents. John really has represented to himself the possible behavioural scenario 'that we should stay at the bus stop' and 'that we should run from it to the local police station.' Also, he has represented to himself a web of more general beliefs and desires, which he has correlated up with these two behavioural scenarios and which make these behavioural scenarios relevant and plausible and sensible solutions to his problem. That is, he has also evaluated these behavioural scenarios in such a way that he can be said truly and literally to decide on one of them for good reasons.

From the above example, we find that John's mind must be able to make use of some medium in terms of which he can represent the behavioural scenarios, beliefs, desires etc. And John's brain must have a language of thought whereby propositional contents of beliefs, decisions, and the other propositional attitudes are first represented, and then operated on or processed in the individual ways which go to form the different propositional attitudes. For the same propositional content, for example 'that there will be rain', could be the information content of two different attitudes. One can believe that there will be rain and hope that there will be rain or believe it but not hope for it.

According to Fodor, the theory of propositional attitudes is required to meet some conditions. There are basically two such conditions. First, a theory of propositional attitudes must explain the parallelism between verbs of propositional attitudes and verbs of sayings. Second, it must explain the opacity of propositional attitudes. By parallelism between verbs of sayings and verbs of propositional attitudes, it can be shown that 'John believes that it is raining' and its corresponding in verb of saying, namely 'John says that it is raining' exhibit isomorphism in syntax, semantics and logical form of structure. The opacity of propositional attitudes is a complex phenomena understood in terms of the following three characteristics. Firstly, statements containing verbs of propositional attitudes are not truth-functions of their components. For example, from the truth of the declarative sentence, 'George Orwell wrote *Animal Farm*', we cannot compute the truth of the statement 'John believes that George Orwell wrote Animal Farm.' Secondly, though the declarations warrant existential generalizations, a statement occurring as the object of verbs of the propositional attitudes does not warrant such existential generalizations. For example, from the truth of the statement 'George Orwell wrote Animal Farm', we can infer: There is George Orwell, who wrote *Animal Farm*. But from this we cannot infer 'There is a George Orwell John believes that George Orwell wrote *Animal Farm*'. The third feature of the opacity of propositional attitudes is that in the case of propositional attitude the principle of substitution fails. The principle of substitution says that given a true statement of identity, one of its terms can be substituted for the other in any true statements where one of the terms of the statement occurs and the resulting statement is true.

## Hypothesis of Language of Thought and Semantic Engines

The representational theory of mind arises with the recognition that thoughts have contents carried by mental representations. For example, John thinks that snow is white. Here, John's mental representation or thought has the mental content that snow is white. As we know that there are different kinds of representations such as pictures, maps, etc, which refer to something. In this case, we are talking about only mental representations. Sententialism distinguishes itself as a version of representationalism by positing that mental representations are themselves linguistic expressions within a 'language of thought'. Moreover, on the other hand, some sententialists point out that the language of thought is just the thinker's spoken language internalized, others identify the language of thought with mentalese that is an unarticulated and internal language in which the computations occur.

Beliefs are true or false. If beliefs are relations to mental representations, then beliefs must have seen relations to representations that have truth-values among their semantic properties. And if sententialism says that mental representations have truth-values, we could readily account for the true valuation of mental representations. Belief plays a central part in reasoning. As well-known, reasoning is a process that attempts to secure new beliefs by exploiting beliefs. Reasoning would preserve truth of belief by being the manipulation of truth-valued sentential representations according to the rules. Thus the sententialist hypothesis is that reasoning consists in formal inferences; it is a process tuned primarily to the structure of mental sentences. Then, reasons are things very much like classical programmed computers.

According to Fodor, thinking is systematic and productive. For example, John wonders whether William is taller than Russell. This implies that John is capable of considering that Russell is shorter than William. More clearly, the fact that John can have some thought entails that he can have certain other semantically related thoughts. Now the question is: How is this systematicity possible?

According to Fodor, "the property of linguistic capacities that I have in mind is one that inheres in the ability to understand and produce sentences. That ability is- as I shall say- systematic: by which I mean the ability to produced/understand some of the sentences is intrinsically connectd to the ability to produced/understand many of the others." For example, suppose that John's thought that William is taller than Russell involves the registration of 'William is taller than Russell.' This mentalese sentence is itself a complex representation containing simpler representations. As complex mental representations, mental sentences result from processes ultimately defined on mentalese words and expressions. So if, John can produce 'William is taller than Russell,' he must have access to 'William', 'Russell' and 'is taller than'. And if he has these mental representations, he is capable of producing 'Russell is shorter than William'. Thus sententialism posits that mental representations are linguistically complex representations whose semantic properties are determined by the semantic properties of their constituents.

Productivity and systematicity run together; if we postulate mechanisms adequate to account for the one, then we get the other automatically. Now the question is: What sort of mechanisms? The sentences of a natural language have a combinatorial semantics. On this view, learning a language is learning a perfectly general procedure for determining the meanings of its lexical elements. Linguistic capacities cannot help but be systematic on this account, because it gives the very same combinatorial mechanisms that determine the meaning of all of the rest. Language expresses thought and thought is as systematic as language is.

To have the thought that William is taller than Russell is *ipso facto* to have access to the thought that Russell is shorter than William. Of course, anybody who is in a position to have one of these thoughts is *ipso facto* in a position to have the other. Thus the language of thought explains the systematicity of thought, which is an essential requirement of the structure of language of thought.

Moreover, in the mind there are two boxes: a belief box and desire box. The 'language of thought' hypothesis is a speculation on the form that storage takes place. Our belief and desires are encoded as sentences. According to Fodor, our sentences that we think are not the English sentences or any sentences in natural languages. Our thinking occurs in a special language, called mentalese. Mentalese is organized into words and sentences. Mentalese words are concepts; mentalese sentences are thoughts. The sentences of Mentalese are stored in a neural medium because patterns of neural activity could develop sentential representations. Words can be configurations of neurons as easily as configurations of chalk marks on the black board.

Fodor's language of thought fits with the multiple realizability arguments because according to LoT, cognition has nothing directly to do with its species-specific neurobiological embodiments, but rather concerns processes operating on the common language of thought. Here, cognition is neural, but cognition is computational in Mentalese.

However, we can imagine a device that could manipulate sentences without regards to their meanings, but it can be done in a way that coincides with the way those sentences would be manipulated by someone who knew their meanings. Such kind of a device, Haugeland calls 'semantic engines'. This device would perfectly mimic the performance of a native speaker, but would do so without relying, as a native speaker would do, on the meanings of the sentence manipulated. Those sentences may express propositions, but the device cares only about their shapes that is 'syntax'. In this way, Haugeland said, "If you take care of syntax, semantics will take care of itself."

Now the question is, is such a device possible? According to Haugeland, not only are semantic engines possible, but they exist. An ordinary computing machine is a semantic engine. We design and programme computers so that they manipulate symbols in accordance with purely syntactic rules. The symbols are meaningful to us but the machines that deploy them care nothing about this. They operate on uninterrupted symbols, but in a way, that honour semantic constraints. The question is: How can syntax mirror semantics? The formal logic is the best example:

$$P \supset Q$$

$$P$$

$$\therefore \quad Q$$

This rule tells us that if we have a particular configuration of symbols, we are permitted to write new symbols. Here what is significant about the *modus ponens* rule is that it is formulated and deployed without regard to semantics. But the rules make sense that is to say, they conform to the semantics of inferences. The concrete example is:

If it is raining, then I shall need an umbrella.

P = It is raining.

Q = I shall need an umbrella.

Thus systems of formal logic mirror this kind of semantic knowledge in rules, and the application of which requires no semantic knowledge. The question is: What has this to do with minds? To explain the human mind by supposing that minds manipulate mental representations, we need sentences in the language of thought. If minds manipulate sentences, then this would seem to require a 'sentence under-stander,' some component of the mind that inputs the symbols. Another job of the mind is to understand sentences in the language of thought. Against this background, it is easier to apply the notion of semantic engines. We have already mentioned that a semantic engine is a device that performs symbol operations in a way that reflects semantic relations holding among those symbols, but does so exclusively by syntactic principles. In the same way, we can also suppose that mind contains mechanisms which understand the meanings of those representations.

Therefore, if the mind is a semantic engine realized by the brain, if the mental operations include the manipulation of symbols, i.e., sentences in the language of thought, then the embodiments of those symbols in the brain need not resemble the symbols we write with pen and paper. They might involve subtle electrical or chemical states. If there is a language of thought, its sentences are invisible from the point of view of an observer examining the microstructure of a brain.

## Computational Theory of Mind

In philosophy, a computational theory of mind names a view that the human mind or the human brain (or both) is an information processing system and that thinking is a form of computing. The theory was proposed in its modern form by Hilary Putnam in 1961, and developed by the MIT philosopher and cognitive scientist Jerry Fodor (who was Putnam's PhD student) in the 1960s, 1970s and 1980s. Despite being vigorously disputed in analytic philosophy in the 1990s (due to work by Putnam himself, John Searle, and others), the view is common in modern cognitive psychology and is presumed by many theorists of evolutionary psychology; in the 2000s and 2010s the view has resurfaced in analytic philosophy (Scheutz 2003, Edelman 2008).

The computational theory of mind holds that the mind is a computation that arises from the brain acting as a computing machine. The theory can be elaborated in many ways, the most popular of which is that the brain is a computer and the mind is the result of the program that the brain runs. A program is the finite description of an algorithm or effective procedure, which prescribes a sequence of discrete actions that produces outputs based only on inputs and the internal states (memory) of the computing machine. For any admissible input, algorithms terminate in a finite number of steps. So the computational theory of mind is the claim that the mind is a computation of a machine (the brain) that derives output representations of the world from input representations and internal memory in a way that is consistent with the theory of computation.

Computational theories of mind are often said to require mental representation because 'input' into a computation comes in the form of symbols or representations of other objects. A computer cannot compute an actual object, but must interpret and represent the object in some form and then compute the representation. The computational theory of mind is related to the representational theory of mind in that they both require that mental states are representations. However the two theories differ in that the representational theory claims that all mental states are representations while the computational theory leaves open that certain mental states, such as pain or depression, may not be representational and therefore may not be suitable for a computational treatment. These non-representational mental states are known as qualia. In Fodor's original views, the computational theory of mind is also related to the language of thought. The language of thought theory allows the mind to process more complex representations with the help of semantics.

## "Computer Metaphor"

Computational theory of mind is not the same as the computer metaphor, comparing the mind to a modern-day digital computer. Computational theory just uses some of the same principles as those found in digital computing.

'Computer' is not meant to mean a modern-day electronic computer. Rather a computer is a symbol manipulator that follows step by step functions to compute input and form output. Alan Turing describes this type of computer in his concept of a Turing Machine.

## Early Proponents

One of the earliest proponents of the computational theory of mind was Thomas Hobbes, who said, "by reasoning, I understand computation. And to compute is to collect the sum of many things added together at the same time, or to know the remainder when one thing has been taken from another. To reason therefore is the same as to add or to subtract." Since Hobbes lived before the contemporary identification of computing with instantiating effective procedures, he cannot be interpreted as explicitly endorsing the computational theory of mind, in the contemporary sense.

## Causal Picture of Thoughts

At the heart of the Computational Theory of Mind is the idea that thoughts are a form of computation, and a computation is by definition a systematic set of laws for the relations among representations. This means that a mental state represents something if and only if there is some causal correlation between the mental state and that particular thing. An example would be seeing dark clouds and thinking "clouds mean rain", where there is a correlation between the thought of the clouds and rain, as the clouds causing rain. This is sometimes known as Natural Meaning. Conversely, there is another side to the causality of thoughts and that is the non-natural representation of thoughts. An example would be seeing a red traffic light and thinking "red means stop", there is nothing about the color red that indicates it represents stopping, and thus is just a convention that has been invented, similar to languages and their abilities to form representations.

## Semantics of Mental States

The computational theory of mind states that the mind functions as a symbolic operator, and that mental representations are symbolic representations; just as the semantics of language are the features of words and sentences that relate to their meaning, the semantics of mental states are those meanings of representations, the definitions of the 'words' of the language of thought. If these basic mental states can have a particular meaning just as words in a language do, then this means that more complex mental states (thoughts) can be created, even if they have never been encountered before. Just as new sentences that are read can be understood even if they have never been encountered before, as long as the basic components are understood, and it is syntactically correct. For example: "I have eaten plum pudding every day of this fortnight." While it's doubtful many have seen this particular configuration of words, nonetheless most readers should be able to glean an understanding of this sentence because it is syntactically correct and the constituent parts are understood.

## Criticism

A range of arguments have been proposed against Computational Theories of Mind.

An early, though indirect, criticism of the Computational Theory of Mind comes from philosopher John Searle. In his thought experiment known as the Chinese room, Searle attempts to refute the claims that artificially intelligent systems can be said to have intentionality and understanding and that these systems, because they can be said to be minds themselves, are sufficient for the study of the human mind. Searle asks us to imagine that there is a man in a room with no way of communicating to anyone or anything outside of the room except for a piece of paper with symbols written on it that is passed under the door. With the paper, the man is to use a series of provided rule books to return paper containing different symbols. Unknown to the man in the room, these symbols are of a Chinese language, and this process generates a conversation that a Chinese speaker outside of the room can actually understand. Searle contends that the man in the room does not understand the Chinese conversation. This is essentially what the computational theory of mind presents us—a model in which the mind simply decodes symbols and outputs more symbols. Searle argues that this is not real understanding or intentionality. Though originally written as a repudiation of the idea that computers work like minds, it is not a stretch to also argue from this position that minds do not work like computers.

Searle has further raised questions about what exactly constitutes a computation:

the wall behind my back is right now implementing the WordStar program, because there is some pattern of molecule movements that is isomorphic with the formal structure of WordStar. But if the wall is implementing WordStar, if it is a big enough wall it is implementing any program, including any program implemented in the brain.

Putnam himself became a prominent critic of computationalism for a variety of reasons, including ones related to Searle's Chinese room arguments, questions of world-word reference relations, and thoughts about the mind-body relationship. Regarding functionalism in particular, Putnam has claimed along lines similar to, but more general than Searle's arguments, that the question of whether the human mind *can* implement computational states is not relevant to the question of the nature of mind, because "every ordinary open system realizes every abstract finite automaton." Computationalists have responded by aiming to develop criteria describing what exactly counts as an implementation.

Roger Penrose has proposed the idea that the human mind does not use a knowably sound calculation procedure to understand and discover mathematical intricacies. This would mean that a normal Turing complete computer would not be able to ascertain certain mathematical truths that human minds can.

## Propositional Attitudes (PAs)

Generally, propositional attitudes (PAs) are the attitudes towards propositions. Our everyday conception of mentality bristles with notions like belief, desire, intention, hope, etc. These attitudes are identified by their propositional contents. For example, a belief that snow is white is identified by the proposition that snow is white. On the other hand, propositional attitudes are pervasive in our descriptive, explanatory, and justificatory practices. we shall show that propositional attitudes are relations between organisms and the external environment.

Before analyzing the above point, now we have to clarify the relationship between intentationality and propositional attitudes. Searle says, "intentionality is that property of many mental states and events by which they are directed at or about or of objects and state of affairs in the world." For example, if I have a belief, it must be a belief that such and such is the case. If I have a fear, it must be a fear of something or that something will occur; similarly, if I have a desire, it must be desired to do something. The above grammar of intentionality shows that the propositional attitudes are intentional. This might be called a grammatical form of intentionality.

According to Fodor, propositional attitudes should be analysed as relations. For example, the verb in a sentence like 'John believes it is raining' expresses a relation between John and something else, and a token of that sentence is true if John stands in the belief-relation to that thing. 'John believes it is raining' is true in virtue of a belief-making relation between John and a token of F (it is raining). It is the complement of a belief-ascriber that determines which internal formula is involved in its truth conditions; in effects 'it is raining' in 'John belief it is raining' functions as an index which picks out F (it's raining) and not, for example, F (elephants have wings).

Moreover, Fodor says, "I am taking seriously the idea that the system of internal representations constitute a (computational) language. Qua language, it presumably has a syntax and a semantics; specifying the language involves saying what the properties are in virtue of which its formulae are well-formed, and what relation(s) obtain between the formulae and things in the (non-linguistic) world. I have no idea what an adequate semantics for a system of internal representations would look like; suffice it that, if propositions come in at all, they come in here." In particular, nothing stops us from specifying semantics for the IRS (Intentional Realism System) by saying that some of its formulae express propositions. If we do say, this, then we can make sense of the notion that propositional attitudes are relations to propositions. Therefore, they are mediated relations to propositions with internal representations doing the mediating.

Folk psychology shows that the propositional attitudes like belief, desire, etc. that are real, and that are part of the mental world of the human beings. Fodor defends folk psychology against the anti-folk psychologists like Dennett and Churchland who maintain that the belief-desire psychology is describable as the human mind is nothing but the brain.

According to P. M. Churchland, mental states are identical with brain states and so, if our mental states are in some sense identical with those states, then we have no reason to refute materialism. Churchland holds that folk psychology which displays propositional attitudes is similar with mathematical physics which displays numerical attitudes. For example, in folk psychology if X fears that P, then X desires that not-P. Similarly, the laws of mathematical physics display a precisely parallel structure; in the latter it is only numerical relations that are being exploited, rather than logical relations. For example, if X has a mass of M, and X suffers a net force of F, then X has an acceleration of F/M. As P. M. Churchland says, "What is taking place in such examples is the same in all cases. The abstract relations holding in the domain of certain abstract objects numbers, or vectors, or propositions-are drawn upon to help us state the empirical regularities that hold between real states and objects, such as between temperatures and pressures, forces and accelerations, interacting momenta, ...and between various types of mental states. The conceptual framework of folk-psychology is exploiting an intellectual strategy that is standard in many of our conceptual endeavours. And just as a theory is neither essentially physical, nor essentially nonphysical, for exploiting numbers and vectors; neither is a theory essentially physical, nor essentially non-physical, for exploiting numbers or vectors; neither is a theory essentially physical, nor essentially non-physical, for exploiting propositions. It remains an empirical question whether the propositional attitudes are ultimately physical in nature. The mere fact that they are PAs (and hence displaying intentionality) entails nothing one way or other."

Dennett argues that we can attribute internal functional states to humans not on the basis of any neurophysiological knowledge, which ordinary people do not have, but on the basis of an observation of how that person behaved in the light of what she or he perceives in his or her environment. Dennett puts it, "In the light of perceptual input and behavioural output we project upon that person certain central (in the hand) functions for that is how our ordinary, everyday psychological explanations operate. Thus it can be said, to attribute an in-the-head or centralist function to some person neither as a result of correctly inferring real internal 'intervening' processes. Rather, we make such attribution as a result of guessing what part, when speaking in a special 'intentional function' way, the brain and central nervous system would have played in the complex production line of perceptual input, central processing, and behaviour output, if it were an 'intentional engine'."

This intentional description can be seen to be the expression of a particular sort of attitudes or stance, which the humans have towards other humans and animals. As Dennett put it, "I wish to examine the concept of system whose behaviour can be- at least sometimes- explained and predicted by relying on ascriptions to the system of beliefs and desires (and hopes, fear, intentions, hunches...). I will call such systems, and predications intentional explanations and predications, in virtue of the intentionlity of the idioms of beliefs and desire (and hope, fear, intention, hunch...)."

In general one can take up the intentional stance in order to explain and predict and so plan or take action. But, we should be clear that Dennett's view differs from Fodor's. As Dennett makes it clear, "the definition of intentionl system I have given does not say that intentional systems really have beliefs and desires, but that one can explain and predict their behaviour by ascribing beliefs and desires to them." Dennett points out there are other stances, besides the intentional, which we take up to things. For example, we might consider a machine from the point of view of its design, that is, we might take up the design stance.

If one knows exactly how a computer or machine is designed, one can predict its designed response to any more one makes by following the computational instructions of the program. For example, the radio engineer's schematic wiring diagrams have symbols for each resistor, capacitor, transistor, etc., each with its tasks to perform and he or she can give a design-stance prediction which are generated by assuming that each element performs its task.

The essential feature of the design stance is that we make predictions solely from knowledge or assumptions about the system's functional design, irrespective of the physical constitution of the particular object. That Dennett calls the physical stance. This attitude or stance is to consider something only in so far as it is made of a certain material or certain types of material which have certain properties. To take up the physical stance to humans is to investigate their psychology or at a more basic level, their physics and chemistry.

However, on the other hand, the core of Dretske's account of intentionality lies in his account of the human brain and its perceptual organs as an information processing system, which in turn is based on 'information theory' developed by the cognitive theorists. This information-processing account of our mental life is purely physicalist in nature. So Dretske's account of intentionality of mental function is materialistic. The information-processing input mechanisms we call the senses, and by treating the brain as an information processor, we can build up an account of intentional states such as knowledge and belief.

Dretske suggests, "The interplay of our human perceptual and cognitive systems is based on this transformation of information from analogue to digital form. It is the successful conversion of information into digital form that constitutes the essence of cognitive activity." Now we have to find out how physical structures which carry information in analogue form can be transformed into physical structures that carry information into digital form, so that this digitalized information becomes a true semantic content and so able to be the content of some mental act such as a belief. For example, humans acquire the concept of red by seeing to red objects, that is, by having his or her visual perception stimulated by red objects. Internally some structures will be selected as the analogue registering structures for 'red stimulation of the visual system' if the person concerned is also exposed to a good number of things which are not red. Such structures become one which has semantic content, and so one which is presenting a utilizable concept of red to the person whose brain contains structures, only when this structure has been made precise and determinate. In this way, it can be shown that the semantic content has nothing to do with the behavioural output of the system. When this semantic content is utilized so as to guide behaviour, it is employed as a map by which the person whose head contains this semantic structure finds his way about the world. Semantic contents become beliefs in so far as they are used as maps or representations to guide output behaviour. Therefore, semantic content becomes a cognitive content when it gains a functional role.

## Intentional Realism

A realistic theory of mind holds intentional realism. Intentional realism is a thesis that the mind is primarily a representational system or that an individual individual's mind is a system whose job it is to deliver representations of the environment for the benefit of the individual whose mind it is. An important problem for intentional realism is to offer an account of how intentional states can be causally related to one another and to the world and to the behaviour of the rational agents. Intentionality is admitted by Fodor as a real feature of the mental representations which can be computationally studied. But, there is distinction between intentional realism and intentional irrealism. The weak sense of representation is intentional irrealism. This thesis claims that the so-called propositional attitudes can be thought of as mental representations of state of affairs. The strong-sense of representationalism which is the claim that the representation properties of propositional attitudes can take us some way towards understanding aspect of conscious experience.

According to intentional irrealism, any utterance of a sentence ascribing a semantic property to an individual's propositional attitude express false properties. The intentional irrealist is also known as an eliminativist materialist according to whom "no mental difference without some physical difference" According to eliminative materialism, "the thesis that our commonsense conception of psychological phenomena constitutes a radically false theory, a theory so fundamentally defective that both the principles and the ontology of that theory will eventually be displaced, rather than smoothly reduced, by complicated neuroscience." This theory says that our folk concepts of propositional attitudes with their purported semantic properties are best compared to such concepts as the physical and chemical concepts of caloric and phlogiston and other alchemical concepts, that is, concept devoid of reference. Therefore, there are no such states as propositional attitudes with semantic properties.

On the other hand, the non-factual version of intentional irrealism is the claim that predicates that are used to refer to semantic properties of an individual's propositional attitude simply do not stand for any genuine properties at all. This version of intentional irrealism has been advocated by Stich. According to him, "such predicates typically do not express properties at all. What this suggests is that there is no such thing as the property of believing that P. The predicate 'is a belief that P' does not express or correspond to a property. If this is right, then we have yet another reason for not thinking of folk psychological beliefs as state tokens, since a state token is the instantiation of a property by an individual during a time interval, and if there is no property, then there can be no state token. It is important to realize, however, that the non-existence of belief properties and belief state tokens do not entail that predicates of the form 'is belief that P' are meaningless or never apply to anything...."

The above, non-factualist interpretation of intentional irrealism is the influential view of Dennett, who has introduced the concept of intentional stance. The intentional stance says that to attribute propositional attitudes to a physical system is not to attribute semantic properties to the system. Rather, it consists in adopting a certain heuristic stance towards it which in turn serves pragmatic goals. According to Dennett, "the decision to adopt the strategy is pragmatic, and is not intrinsically right or wrong." Thus the intentional stance view says that semantic properties of an individual's propositional attitudes arise from the stance taken towards the individual by an observer or an interpreter.

Intentional realism is different from both error-theorism and non-factualist irrealism because intentional realist is committed to the view that the semantic properties of an individual's propositional attitudes are genuine properties of the individual's brain. According to P. Jacob, we can ascribe three theses to intentional realism:

i. The semantic properties of an individual's propositional attitudes are genuine of the individual.

ii. The semantic properties of an individual's utterance are derivative from the semantic properties of his or her propositional attitudes.

iii. The semantic properties of an individual's propositional attitudes must contribute to the production of the individual's intentional behaviour.

If semantic properties are genuine properties, then having a mind must make a causal difference. Minded systems must be able to do things which systems lacking a mind must be unable to do. And if having a mind did not make a causal difference, then the question is, what would it do to have a mind? Therefore, states of mind must be causes. But the fact that minds can occupy states with semantic properties can explain why systems having a mind can do things which systems without a mind cannot do. This is the problem of mental causation. Thus Fodor says, "I suppose that sooner or later the physicists will complete the catalogue they have been compiling of the ultimate and irreducible properties of things. When they do, the likes of spin, charm, and change will perhaps appear upon their list. But aboutness surely would not; intentionality simply does not go that deep. It's hard to see, in face of this consideration, how one can be Realist about intentionality without also being, to some extent or other, a Reductionist. If the semantic and the intentional are real properties of things, it must be in virtue of their identity with (or may be supervenient on) properties that are themselves neither intentional nor semantic. If aboutness is real, it must be really be something else."

The problem raised by Fodor shows the intentional realist's dilemma. On the other hand, that intentional realist is a physicalist. The mind must be complex physical systems. On the other hand, he is a realist about minds. It is the view of the anti-reductionist that mind passes semantic properties which must make a causal difference. The issue is closely related to the issue of reducibility of a system's semantic properties to its non-semantic properties. According to Jacob, there are two ways one can think about reduction. For example, water turned out to be identical to H2O molecules; and genes turn out be nothing but DNA molecules. Such identities are nomic in the sense that what is claimed is that nothing would be water unless it were composed of H2O molecules. On the other hand, semantic properties are reduced to non-semantic properties, on the ground that the latter provide necessary and sufficient non-semantic conditions for the possession of semantic properties. The intentional realist like Fodor tries to bridge the gap between semantic properties and non-semantic properties. This is the main dilemma of intentional realist.

## References


- Augusto, Luis M. (2014). "Unconscious representations 2: Towards an integrated cognitive architecture.". Axiomathes. 24: 19–43. doi:10.1007/s10516-012-9207-y
- Malle, Betram F; Knobe, Joshua (Mar 1997). "The folk concept of intentionality". Journal of Experimental Social Psychology. 33 (2): 101–121. doi:10.1006/jesp.1996.1314
- Fletcher, G. (1995). The scientific credibility of folk psychology. Hilsdale, NJ: Lawrence Erlbaum Associates Inc. ISBN 0805815708. OCLC 476199418
- Arico, Adam (2010). "Folk psychology, consciousness, and context effects". Review of Philosophy and Psychology. 1 (3): 317–393. doi:10.1007/s13164-010-0029-9. Retrieved 10 March 2012
- Kashima, Yoshihisa; McKintyre, Allison; Clifford, Paul (1 April 2000). "The category of the mind: Folk psychology of belief, desire, and intention. Author". Asian Journal of Social Psychology. 1. 1 (3): 289–313. doi:10.1111/1467-839X.00019
- Marr, David (2010). Vision. A Computational Investigation into the Human Representation and Processing of Visual Information. The MIT Press. ISBN 978-0262514620
- Boonzaier, A.; McClure, J.; Sutton, R. M. (2005). "Distinguishing the effects of beliefs and preconditions: The folk psychology of goals and actions". European Journal of Social Psychology. 35 (6): 725–740. doi:10.1002/ejsp.280
- Hilton, Denis J. (Jan 1990). "Conversational processes and causal explanation". Psychological Bulletin. 107 (1): 65–81. doi:10.1037/0033-2909.107.1.65

# 3

# Limitations of Artificial Intelligence

Philosophy of artificial intelligence questions the progress of artificial intelligence. It questions the advances in artificial intelligence and its ability to gain consciousness and develop a mind similar to humans. This section provides a plethora of interdisciplinary topics for better comprehension of artificial intelligence.

## Artifical Intelligence

It is difficult to give a precise definition of artificial intelligence. Some recent artificial intelligence scientists have attempted to define artificial intelligence (in short AI) in various ways.

According to Haugeland, artificial intelligence is, "the exciting new effort to make computers think ... machines with minds, in the full and literal sense"[1] For Bellman, it is "the automation of activities that we associates with human thinking, activities such as decision making, problem solving, learning..." Charniak and McDermett define AI as "the study of mental faculties through the use of computational model." And for Winston, it is "the study of the computations that make it possible to perceive, reason and act." AI, for Kurzweil is "the art of creating machines that perform functions that require intelligence when performed by people." Rich and Knight say that AI is "The study of how to make computers thinks at which, at the movement, people are better." For Schalkoff, AI is "a field of study that seeks to explain and emulate intelligent behavior in terms of computational process." Luger and Stubblefield hold it to be "the branch of computer science that is concerned with the automation of intelligent behaviour."

Let us look at all the definitions from different angles. Haugeland and Bellman point out that artificial intelligence is concerned with thought process and reasoning. They have explained the mind as a machine that is completely associated with human thinking. That is to say, computers do think. But Schalkoff, Luger and Stubblefield are concerned with the behavioural aspects of systems. For them, computers behave as intelligently as human beings. Moreover, Kurzweil, Rich and Knight are concerned with or measure success in terms of human performance. For them, artificial intelligence can be attributed to machines, but it belongs basically to the human mind. At last, Charniak, McDermett and Winston are concerned with an ideal intelligence. They explain the mental faculties through the use of computational models.

To sum up, all the definitions of AI can be organized into four categories. They are as follows:

i. Systems that think like humans.

ii. Systems that act like humans.

iii. Systems that think rationally.

iv. Systems that act rationally.

Now, we have to look at each aspect in detail.

The Turing test, as named after Alan Turing, was designed to provide a satisfactory operational definition of intelligence. Turing defined that intelligent behavior as the ability to achieve human-level performance in all cognitive tasks to fool an interrogator. In his '*Computing Machinery and Intelligence*', Turing says the new form of the problem can be described in terms of a game which we call the '*imitation game*.' It is played by a man (A), a woman (B), and an interrogator (C) who may be of either sex. The interrogator stays in a room apart from the other two. The object of the game for the interrogator is to determine which of the other two is the man and which is the woman. He or She knows them by labels X and Y, and at the end of the game, he or she says, either 'X is A and Y is B' or 'X is B and Y is A.' The interrogator is allowed to put questions to A and B.

Thus, C: will X please tell me the length of his or her hair?

Now suppose X is actually A, then A must answer to the question. It is A's object in the game to try to cause C to make the wrong identification. His or her answer might therefore, be 'my hair is singled, and the longest strands are about nine inches long.'

However, because the tones of voice may not help the interrogator, the answer should be written or better still be typewritten. The ideal arrangement is to have a tele-printer for perfect communication. Alternatively, an intermediary can repeat the questions and answers. The object of the game for the second player (B) is to help the interrogator. The best strategy for her is probably to give truthful answers. She can add to her answer such things as 'I am the woman, do not listen to him,' but it is of no avail as the man can make similar remark.

## Acting Humanly: Turing Machine Approach

Turing's answers to these questions are more or less summed up in the following passage; "I believe that in about fifty years time it will be possible to program computers, with a storage capacity of about 10, to make them play the imitation game so well that an average interrogator will not have more than 70 percent chance of making the right identification after five minutes of questioning."

What Turing had predicted at that time, now is a fact that the machine or the computer can imitate human behaviuor. It should be pointed out that Turing's beliefs about the capabilities and capacities of machines are not limited to such activities as playing the imitation game as successfully as human beings. Roughly speaking, the test Turing proposed is that the computer should be interrogated in the place of human beings.

Turing's test deliberately avoided direct physical interaction between the interrogator and the computer, because physical limitation of a person is unnecessary for intelligence. However, the so-called Turing test includes a-video signal so that the interrogator can test the subject's perceptual abilities. In order to pass through total Turing test, the computer will need computer vision to perceive objects and robotics to move them.

Again, the issue of acting like a human comes up primarily when artificial intelligence programs have to interact with people, as when expert system explains how it came to its diagnosis, or a natural language processing system has a dialogue with a user. These programs must behave according to certain normal covertness of human interaction in order to make them understood. The Turing test shows that machines can interact with human beings the way human beings interact amongst themselves. That is to say that machines can behave the way the human beings do.

## Turing Machine

Automata theory

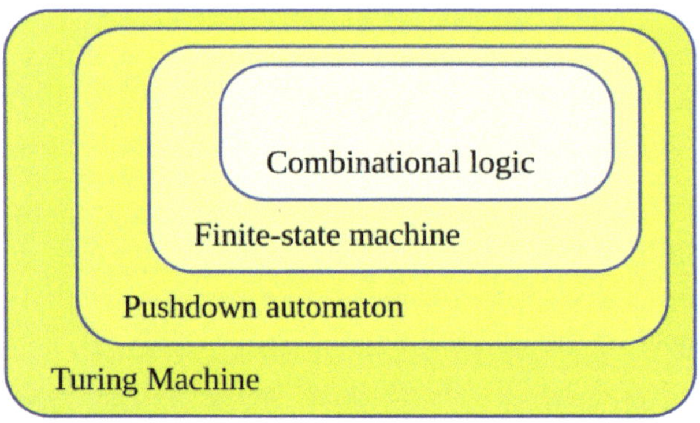


Classes of automata (Clicking on the text will take you to an article on that subject)

A Turing machine is an abstract machine that manipulates symbols on a strip of tape according to a table of rules; to be more exact, it is a mathematical model of computation that defines such a device. Despite the model's simplicity, given any computer algorithm, a Turing machine can be constructed that is capable of simulating that algorithm's logic.

The machine operates on an infinite memory tape divided into discrete *cells*. The machine positions its *head* over a cell and "reads" (scans) the symbol there. Then, as per the symbol and its present place in a *finite table* of user-specified instructions, the machine (i) writes a symbol (e.g. a digit or a letter from a finite alphabet) in the cell (some models allowing symbol erasure and/or no writing), then (ii) either moves the tape one cell left or right (some models allow no motion, some models move the head), then (iii) (as determined by the observed symbol and the machine's place in the table) either proceeds to a subsequent instruction or halts the computation.

The Turing machine was invented in 1936 by Alan Turing, who called it an *a-machine* (automatic machine). With this model, Turing was able to answer two questions in the negative: (1) Does a machine exist that can determine whether any arbitrary machine on its tape is "circular" (e.g. freezes, or fails to continue its computational task); similarly, (2) does a machine exist that can determine whether any arbitrary machine on its tape ever prints a given symbol. Thus by providing a mathematical description of a very simple device capable of arbitrary computations, he was able to prove properties of computation in general—and in particular, the uncomputability of the Entscheidungsproblem ("decision problem").

Thus, Turing machines prove fundamental limitations on the power of mechanical computation. While they can express arbitrary computations, their minimalistic design makes them unsuitable for computation in practice: real-world computers are based on different designs that, unlike Turing machines, use random-access memory.

Turing completeness is the ability for a system of instructions to simulate a Turing machine. A programming language that is Turing complete is theoretically capable of expressing all tasks accomplishable by computers; nearly all programming languages are Turing complete if the limitations of finite memory are ignored.

## Overview

A Turing machine is a general example of a CPU that controls all data manipulation done by a computer, with the canonical machine using sequential memory to store data. More specifically, it is a machine (automaton) capable of enumerating some arbitrary subset of valid strings of an alphabet; these strings are part of a recursively enumerable set.

Assuming a black box, the Turing machine cannot know whether it will eventually enumerate any one specific string of the subset with a given program. This is due to the fact that the halting problem is unsolvable, which has major implications for the theoretical limits of computing.

The Turing machine is capable of processing an unrestricted grammar, which further implies that it is capable of robustly evaluating first-order logic in an infinite number of ways. This is famously demonstrated through lambda calculus.

A Turing machine that is able to simulate any other Turing machine is called a universal Turing machine (UTM, or simply a universal machine). A more mathematically oriented definition with a similar "universal" nature was introduced by Alonzo Church, whose work on lambda calculus intertwined with Turing's in a formal theory of computation known as the Church–Turing thesis. The thesis states that Turing machines indeed capture the informal notion of effective methods in logic and mathematics, and provide a precise definition of an algorithm or "mechanical procedure". Studying their abstract properties yields many insights into computer science and complexity theory.

## Physical Description

In his 1948 essay, "Intelligent Machinery", Turing wrote that his machine consisted of:

> ...an unlimited memory capacity obtained in the form of an infinite tape marked out into squares, on each of which a symbol could be printed. At any moment there is one symbol in the machine; it is called the scanned symbol. The machine can alter the scanned symbol, and its behavior is in part determined by that symbol, but the symbols on the tape elsewhere do not affect the behavior of the machine. However, the tape can be moved back and forth through the machine, this being one of the elementary operations of the machine. Any symbol on the tape may therefore eventually have an innings. (Turing 1948, p. 3)

## Informal Description

The Turing machine mathematically models a machine that mechanically operates on a tape. On this tape are symbols, which the machine can read and write, one at a time, using a tape head. Operation is fully determined by a finite set of elementary instructions such as "in state 42, if the symbol seen is 0, write a 1; if the symbol seen is 1, change into state 17; in state 17, if the symbol seen is 0, write a 1 and change to state 6;" etc. In the original article ("On computable numbers, with an application to the Entscheidungsproblem"), Turing imagines not a mechanism, but a person whom he calls the "computer", who executes these deterministic mechanical rules slavishly (or as Turing puts it, "in a desultory manner").

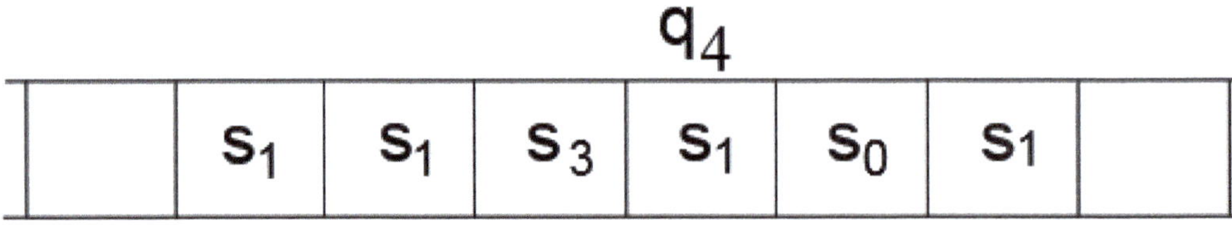


The head is always over a particular square of the tape; only a finite stretch of squares is shown. The instruction to be performed ($q_4$) is shown over the scanned square. (Drawing after Kleene (1952))

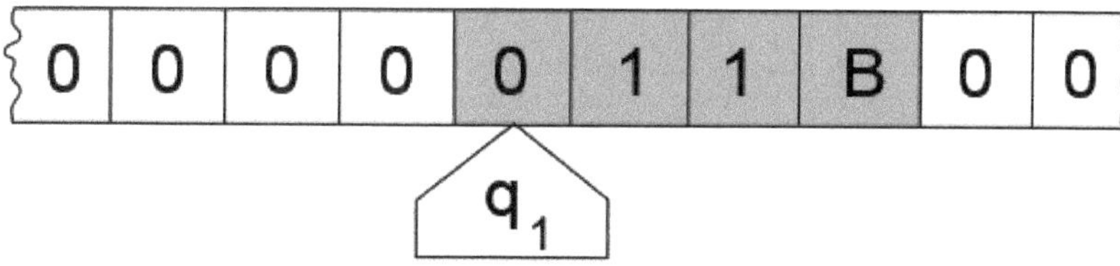


Here, the internal state ($q_1$) is shown inside the head, and the illustration describes the tape as being infinite and pre-filled with "0", the symbol serving as blank. The system's full state (its *complete configuration*) consists of the internal state, any non-blank symbols on the tape (in this illustration "11B"), and the position of the head relative to those symbols including blanks, i.e. "011B".

More precisely, a Turing machine consists of:

1. A tape divided into cells, one next to the other. Each cell contains a symbol from some finite alphabet. The alphabet contains a special blank symbol (here written as '0') and one or more other symbols. The tape is assumed to be arbitrarily extendable to the left and to the right, i.e., the Turing machine is always supplied with as much tape as it needs for its computation. Cells that have not been written before are assumed to be filled with the blank symbol. In some models the tape has a left end marked with a special symbol; the tape extends or is indefinitely extensible to the right.

2. A head that can read and write symbols on the tape and move the tape left and right one (and only one) cell at a time. In some models the head moves and the tape is stationary.

3. A state register that stores the state of the Turing machine, one of finitely many. Among these is the special start state with which the state register is initialized. These states, writes Turing, replace the "state of mind" a person performing computations would ordinarily be in.

4. A finite table of instructions that, given the state($q_i$) the machine is currently in and the symbol($a_j$) it is reading on the tape (symbol currently under the head), tells the machine to do the following in sequence (for the 5-tuple models):

   - Either erase or write a symbol (replacing $a_j$ with $a_{j1}$), *and then*

   - Move the head (which is described by $d_k$ and can have values: 'L' for one step left *or* 'R' for one step right *or* 'N' for staying in the same place), *and then*

   - Assume the same or a *new state* as prescribed (go to state $q_{i1}$).

   In the 4-tuple models, erasing or writing a symbol ($a_{j1}$) and moving the head left or right ($d_k$) are specified as separate instructions. Specifically, the table tells the machine to (ia) erase or write a symbol *or* (ib) move the head left or right, *and then* (ii) assume the same or a new state as prescribed, but not both actions (ia) and (ib) in the same instruction. In some models, if there is no entry in the table for the current combination of symbol and state then the machine will halt; other models require all entries to be filled.

Note that every part of the machine (i.e. its state, symbol-collections, and used tape at any given time) and its actions (such as printing, erasing and tape motion) is *finite*, *discrete* and *distinguishable*; it is the unlimited amount of tape and runtime that gives it an unbounded amount of storage space.

## Formal Definition

Following Hopcroft and Ullman (1979, p. 148), a (one-tape) Turing machine can be formally defined as a 7-tuple $M = \langle Q, \Gamma, b, \Sigma, \delta, q_0, F \rangle$ where

- $Q$ is a finite, non-empty set of *states*
- $\Gamma$ is a finite, non-empty set of *tape alphabet symbols*
- $b \in \Gamma$ is the *blank symbol* (the only symbol allowed to occur on the tape infinitely often at any step during the computation)
- $\Sigma \subseteq \Gamma \setminus \{b\}$ is the set of *input symbols*
- $\delta : (Q \setminus F) \times \Gamma \rightarrow Q \times \Gamma \times \{L, R\}$ is a partial function called the *transition function*, where L is left shift, R is right shift. (A relatively uncommon variant allows "no shift", say N, as a third element of the latter set.) If $\delta$ is not defined on the current state and the current tape symbol, then the machine halts.
- $q_0 \in Q$ is the *initial state*
- $F \subseteq Q$ is the set of *final* or *accepting states*. The initial tape contents is said to be *accepted* by $M$ if it eventually halts in a state from $F$.

Anything that operates according to these specifications is a Turing machine.

The 7-tuple for the 3-state busy beaver looks like this:

- $Q = \{\mathrm{A}, \mathrm{B}, \mathrm{C}, \mathrm{HALT}\}$
- $\Gamma = \{0, 1\}$
- $b = 0$ ("blank")
- $\Sigma = \{1\}$
- $q_0 = \mathrm{A}$ (the initial state)
- $F = \{\mathrm{HALT}\}$

- $\delta$ = see state-table below

Initially all tape cells are marked with 0.

| State table for 3 state, 2 symbol busy beaver | | | | | | | | | |
|---|---|---|---|---|---|---|---|---|---|
| **Tape symbol** | **Current state A** | | | **Current state B** | | | **Current state C** | | |
| | Write symbol | Move tape | Next state | Write symbol | Move tape | Next state | Write symbol | Move tape | Next state |
| 0 | 1 | R | **B** | 1 | L | **A** | 1 | L | **B** |
| 1 | 1 | L | **C** | 1 | R | **B** | 1 | R | **HALT** |

## Additional Details Required to Visualize or Implement Turing Machines

In the words of van Emde Boas (1990), p. 6: "The set-theoretical object [his formal seven-tuple description similar to the above] provides only partial information on how the machine will behave and what its computations will look like."

For instance,

- There will need to be many decisions on what the symbols actually look like, and a failproof way of reading and writing symbols indefinitely.
- The shift left and shift right operations may shift the tape head across the tape, but when actually building a Turing machine it is more practical to make the tape slide back and forth under the head instead.
- The tape can be finite, and automatically extended with blanks as needed (which is closest to the mathematical definition), but it is more common to think of it as stretching infinitely at both ends and being pre-filled with blanks except on the explicitly given finite fragment the tape head is on. (This is, of course, not implementable in practice.) The tape *cannot* be fixed in length, since that would not correspond to the given definition and would seriously limit the range of computations the machine can perform to those of a linear bounded automaton.

## Alternative Definitions

Definitions in literature sometimes differ slightly, to make arguments or proofs easier or clearer, but this is always done in such a way that the resulting machine has the same computational power. For example, changing the set $\{L,R\}$ to $\{L,R,N\}$, where $N$ ("None" or "No-operation") would allow the machine to stay on the same tape cell instead of moving left or right, does not increase the machine's computational power.

The most common convention represents each "Turing instruction" in a "Turing table" by one of nine 5-tuples, per the convention of Turing/Davis (Turing (1936) in *The Undecidable*, p. 126-127 and Davis (2000) p. 152):

(definition 1): ($q_i$, $S_j$, $S_k$/E/N, L/R/N, $q_m$)

( current state $q_i$ , symbol scanned $S_j$ , print symbol $S_k$/erase E/none N , move_tape_one_square left L/right R/none N , new state $q_m$ )

Other authors (Minsky (1967) p. 119, Hopcroft and Ullman (1979) p. 158, Stone (1972) p. 9) adopt a different convention, with new state $q_m$ listed immediately after the scanned symbol $S_j$:

(definition 2): ($q_i$, $S_j$, $q_m$, $S_k$/E/N, L/R/N)

( current state $q_i$ , symbol scanned $S_j$ , new state $q_m$ , print symbol $S_k$/erase E/none N , move_tape_one_square left L/right R/none N )

For the remainder of this article "definition 1" (the Turing/Davis convention) will be used.

| Example: state table for the 3-state 2-symbol busy beaver reduced to 5-tuples | | | | | |
|---|---|---|---|---|---|
| **Current state** | **Scanned symbol** | **Print symbol** | **Move tape** | **Final (i.e. next) state** | **5-tuples** |
| **A** | 0 | 1 | R | **B** | (**A**, 0, 1, R, **B**) |
| **A** | 1 | 1 | L | **C** | (**A**, 1, 1, L, **C**) |
| **B** | 0 | 1 | L | **A** | (**B**, 0, 1, L, **A**) |
| **B** | 1 | 1 | R | **B** | (**B**, 1, 1, R, **B**) |
| **C** | 0 | 1 | L | **B** | (**C**, 0, 1, L, **B**) |
| **C** | 1 | 1 | N | **H** | (**C**, 1, 1, N, **H**) |

In the following table, Turing's original model allowed only the first three lines that he called N1, N2, N3 (cf. Turing in *The Undecidable*, p. 126). He allowed for erasure of the "scanned square" by naming a 0th symbol $S_0$ = "erase" or "blank", etc. However, he did not allow for non-printing, so every instruction-line includes "print symbol $S_k$" or "erase" (cf. footnote 12 in Post (1947), *The Undecidable*, p. 300). The abbreviations are Turing's (*The Undecidable*, p. 119). Subsequent to Turing's original paper in 1936–1937, machine-models have allowed all nine possible types of five-tuples:

| | **Current m-configuration (Turing state)** | **Tape symbol** | **Print-operation** | **Tape-motion** | **Final m-configuration (Turing state)** | **5-tuple** | **5-tuple comments** | **4-tuple** |
|---|---|---|---|---|---|---|---|---|
| N1 | $q_i$ | $S_j$ | Print($S_k$) | Left L | $q_m$ | ($q_i$, $S_j$, $S_k$, L, $q_m$) | "blank" = $S_0$, 1=$S_1$, etc. | |

| | | | | | | | | |
|---|---|---|---|---|---|---|---|---|
| N2 | $q_i$ | $S_j$ | Print($S_k$) | Right R | $q_m$ | ($q_i$, $S_j$, $S_k$, R, $q_m$) | "blank" = $S_o$, 1=$S_1$, etc. | |
| N3 | $q_i$ | $S_j$ | Print($S_k$) | None N | $q_m$ | ($q_i$, $S_j$, $S_k$, N, $q_m$) | "blank" = $S_o$, 1=$S_1$, etc. | ($q_i$, $S_j$, $S_k$, $q_m$) |
| 4 | $q_i$ | $S_j$ | None N | Left L | $q_m$ | ($q_i$, $S_j$, N, L, $q_m$) | | ($q_i$, $S_j$, L, $q_m$) |
| 5 | $q_i$ | $S_j$ | None N | Right R | $q_m$ | ($q_i$, $S_j$, N, R, $q_m$) | | ($q_i$, $S_j$, R, $q_m$) |
| 6 | $q_i$ | $S_j$ | None N | None N | $q_m$ | ($q_i$, $S_j$, N, N, $q_m$) | Direct "jump" | ($q_i$, $S_j$, N, $q_m$) |
| 7 | $q_i$ | $S_j$ | Erase | Left L | $q_m$ | ($q_i$, $S_j$, E, L, $q_m$) | | |
| 8 | $q_i$ | $S_j$ | Erase | Right R | $q_m$ | ($q_i$, $S_j$, E, R, $q_m$) | | |
| 9 | $q_i$ | $S_j$ | Erase | None N | $q_m$ | ($q_i$, $S_j$, E, N, $q_m$) | | ($q_i$, $S_j$, E, $q_m$) |

Any Turing table (list of instructions) can be constructed from the above nine 5-tuples. For technical reasons, the three non-printing or "N" instructions (4, 5, 6) can usually be dispensed with.

Less frequently the use of 4-tuples are encountered: these represent a further atomization of the Turing instructions (cf. Post (1947), Boolos & Jeffrey (1974, 1999), Davis-Sigal-Weyuker (1994)).

## The "State"

The word "state" used in context of Turing machines can be a source of confusion, as it can mean two things. Most commentators after Turing have used "state" to mean the name/designator of the current instruction to be performed—i.e. the contents of the state register. But Turing (1936) made a strong distinction between a record of what he called the machine's "m-configuration", (its internal state) and the machine's (or person's) "state of progress" through the computation - the current state of the total system. What Turing called "the state formula" includes both the current instruction and *all* the symbols on the tape:

Thus the state of progress of the computation at any stage is completely determined by the note of instructions and the symbols on the tape. That is, the state of the system may be described by a single expression (sequence of symbols) consisting of the symbols on the tape followed by Δ (which we suppose not to appear elsewhere) and then by the note of instructions. This expression is called the 'state formula'.

—*The Undecidable, pp. 139–140, emphasis added*

Earlier in his paper Turing carried this even further: he gives an example where he placed a symbol of the current "m-configuration"—the instruction's label—beneath the scanned square, together with all the symbols on the tape (*The Undecidable*, p. 121); this he calls "the *complete configuration*" (*The Undecidable*, p. 118). To print the "complete configuration" on one line, he places the state-label/m-configuration to the *left* of the scanned symbol.

A variant of this is seen in Kleene (1952) where Kleene shows how to write the Gödel number of a machine's "situation": he places the "m-configuration" symbol $q_4$ over the scanned square in roughly the center of the 6 non-blank squares on the tape and puts it to the *right* of the scanned square. But Kleene refers to "$q_4$" itself as "the machine state" (Kleene, p. 374-375). Hopcroft and Ullman call this composite the "instantaneous description" and follow the Turing convention of putting the "current state" (instruction-label, m-configuration) to the *left* of the scanned symbol (p. 149).

This means: after three moves the tape has ... 000110000 ... on it, the head is scanning the right-most 1, and the state is A. Blanks (in this case represented by "0"s) can be part of the total state as shown here: B01; the tape has a single 1 on it, but the head is scanning the 0 ("blank") to its left and the state is B.

"State" in the context of Turing machines should be clarified as to which is being described: (*i*) the current instruction, or (*ii*) the list of symbols on the tape together with the current instruction, or (*iii*) the list of symbols on the tape together with the current instruction placed to the left of the scanned symbol or to the right of the scanned symbol.

Turing's biographer Andrew Hodges (1983: 107) has noted and discussed this confusion.

## Turing Machine "State"

| The table for the 3-state busy beaver ("P" = print/write a "1") | | | | | | | | | |
|---|---|---|---|---|---|---|---|---|---|
| **Tape symbol** | **Current state A** | | | **Current state B** | | | **Current state C** | | |
| | Write symbol | Move tape | Next state | Write symbol | Move tape | Next state | Write symbol | Move tape | Next state |
| 0 | P | R | **B** | P | L | **A** | P | L | **B** |
| 1 | P | L | **C** | P | R | **B** | P | R | **HALT** |

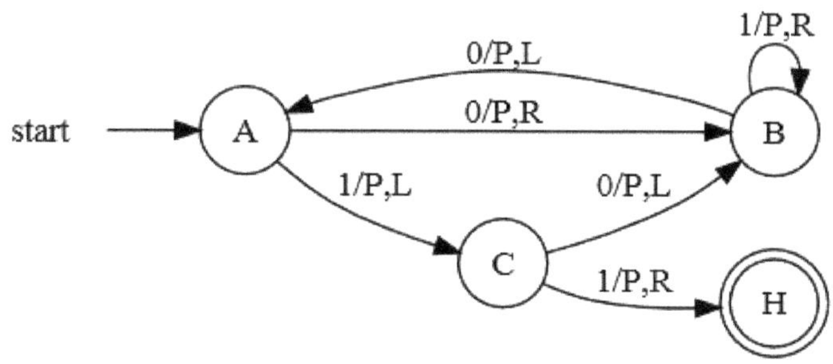


The "3-state busy beaver" Turing machine in a finite state representation. Each circle represents a "state" of the TABLE—an "m-configuration" or "instruction". "Direction" of a state *transition* is shown by an arrow. The label (e.g.. 0/P,R) near the outgoing state (at the "tail" of the arrow) specifies the scanned symbol that causes a particular transition (e.g. 0) followed by a slash /, followed by the subsequent "behaviors" of the machine, e.g. "P Print" then move tape "R Right". No general accepted format exists. The convention shown is after McClusky (1965), Booth (1967), Hill, and Peterson (1974).

To the right: the above TABLE as expressed as a "state transition".

Usually large TABLES are better left as tables. They are more readily simulated by computer in tabular form. However, certain concepts—e.g. machines with "reset" states and machines with repeating patterns (cf. Hill and Peterson p. 244ff)—can be more readily seen when viewed as a drawing.

Whether a drawing represents an improvement on its TABLE must be decided by the reader for the particular context.

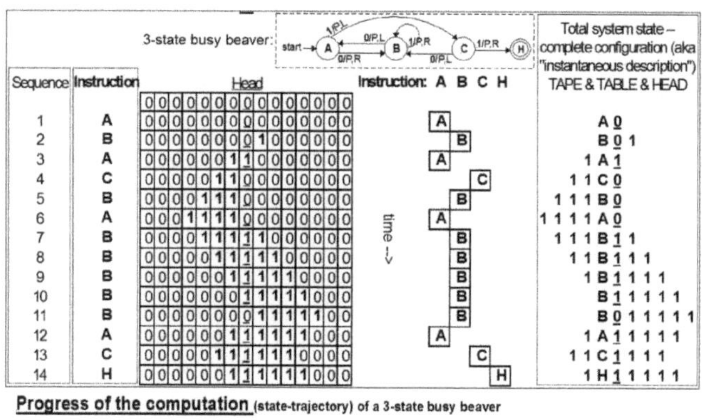


The evolution of the busy-beaver's computation starts at the top and proceeds to the bottom.

The reader should again be cautioned that such diagrams represent a snapshot of their TABLE frozen in time, *not* the course ("trajectory") of a computation *through* time and/or space. While every time the busy beaver machine "runs" it will always follow the same state-trajectory, this is not true for the "copy" machine that can be provided with variable input "parameters".

The diagram "Progress of the computation" shows the 3-state busy beaver's "state" (instruction) progress through its computation from start to finish. On the far right is the

Turing "complete configuration" (Kleene "situation", Hopcroft–Ullman "instantaneous description") at each step. If the machine were to be stopped and cleared to blank both the "state register" and entire tape, these "configurations" could be used to rekindle a computation anywhere in its progress.

## Models Equivalent to the Turing Machine Model

Many machines that might be thought to have more computational capability than a simple universal Turing machine can be shown to have no more power (Hopcroft and Ullman p. 159, cf. Minsky (1967)). They might compute faster, perhaps, or use less memory, or their instruction set might be smaller, but they cannot compute more powerfully (i.e. more mathematical functions). (Recall that the Church–Turing thesis *hypothesizes* this to be true for any kind of machine: that anything that can be "computed" can be computed by some Turing machine.)

A Turing machine is equivalent to a pushdown automaton that has been made more flexible and concise by relaxing the last-in-first-out requirement of its stack.

At the other extreme, some very simple models turn out to be Turing-equivalent, i.e. to have the same computational power as the Turing machine model.

Common equivalent models are the multi-tape Turing machine, multi-track Turing machine, machines with input and output, and the *non-deterministic* Turing machine (NDTM) as opposed to the *deterministic* Turing machine (DTM) for which the action table has at most one entry for each combination of symbol and state.

Read-only, right-moving Turing machines are equivalent to NDFAs (as well as DFAs by conversion using the NDFA to DFA conversion algorithm).

For practical and didactical intentions the equivalent register machine can be used as a usual assembly programming language.

## Choice c-machines, Oracle o-machines

Early in his paper (1936) Turing makes a distinction between an "automatic machine"—its "motion ... completely determined by the configuration" and a "choice machine":

> ...whose motion is only partially determined by the configuration ... When such a machine reaches one of these ambiguous configurations, it cannot go on until some arbitrary choice has been made by an external operator. This would be the case if we were using machines to deal with axiomatic systems.
>
> —*The Undecidable, p. 118*

Turing (1936) does not elaborate further except in a footnote in which he describes how to use an a-machine to "find all the provable formulae of the [Hilbert] calculus" rather than use a choice machine. He "suppose[s] that the choices are always between two possibilities 0 and 1. Each proof will then be determined by a sequence of choices $i_1$, $i_2$, ..., $i_n$ ($i_1$ = 0 or 1, $i_2$ = 0 or 1, ..., $i_n$ = 0 or 1), and hence the number $2^n + i_1 2^{n-1} + i_2 2^{n-2} + ... + i_n$ completely determines the proof. The automatic machine carries out successively proof 1, proof 2, proof 3, ..."

This is indeed the technique by which a deterministic (i.e. a-) Turing machine can be used to mimic the action of a nondeterministic Turing machine; Turing solved the matter in a footnote and appears to dismiss it from further consideration.

An oracle machine or o-machine is a Turing a-machine that pauses its computation at state "**o**" while, to complete its calculation, it "awaits the decision" of "the oracle"—an unspecified entity "apart from saying that it cannot be a machine" (Turing (1939), *The Undecidable*, p. 166–168). The concept is now actively used by mathematicians.

## Universal Turing Machines

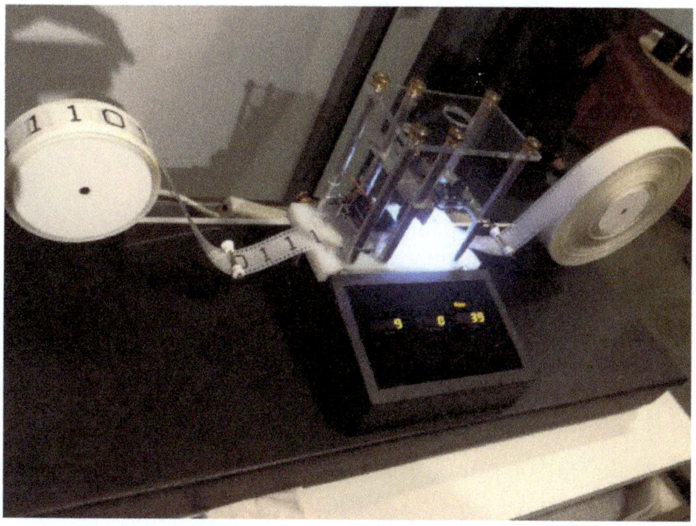

An implementation of a Turing machine

As Turing wrote in *The Undecidable*, p. 128 (italics added):

It is possible to invent a *single machine* which can be used to compute *any* computable sequence. If this machine **U** is supplied with the tape on the beginning of which is written the string of quintuples separated by semicolons of some computing machine M, then U will compute the same sequence as M.

This finding is now taken for granted, but at the time (1936) it was considered astonishing. The model of computation that Turing called his "universal machine"—"**U**" for short—is considered by some (cf. Davis (2000)) to have been the fundamental theoretical breakthrough that led to the notion of the stored-program computer.

Turing's paper ... contains, in essence, the invention of the modern computer and some of the programming techniques that accompanied it.

—*Minsky (1967), p. 104*

In terms of computational complexity, a multi-tape universal Turing machine need only be slower by logarithmic factor compared to the machines it simulates. This result was obtained in 1966 by F. C. Hennie and R. E. Stearns. (Arora and Barak, 2009, theorem 1.9)

## Comparison with Real Machines

A Turing machine realisation in Lego

It is often said that Turing machines, unlike simpler automata, are as powerful as real machines, and are able to execute any operation that a real program can. What is neglected in this statement is that, because a real machine can only have a finite number of *configurations*, this "real machine" is really nothing but a linear bounded automaton. On the other hand, Turing machines are equivalent to machines that have an unlimited amount of storage space for their computations. However, Turing machines are not intended to model computers, but rather they are intended to model computation itself. Historically, computers, which compute only on their (fixed) internal storage, were developed only later.

There are a number of ways to explain why Turing machines are useful models of real computers:

1. Anything a real computer can compute, a Turing machine can also compute. For example: "A Turing machine can simulate any type of subroutine found in programming languages, including recursive procedures and any of the known parameter-passing mechanisms" (Hopcroft and Ullman p. 157). A large enough FSA can also model any real computer, disregarding IO. Thus, a statement about the limitations of Turing machines will also apply to real computers.

2. The difference lies only with the ability of a Turing machine to manipulate an unbounded amount of data. However, given a finite amount of time, a Turing machine (like a real machine) can only manipulate a finite amount of data.
3. Like a Turing machine, a real machine can have its storage space enlarged as needed, by acquiring more disks or other storage media. If the supply of these runs short, the Turing machine may become less useful as a model. But the fact is that neither Turing machines nor real machines need astronomical amounts of storage space in order to perform useful computation. The processing time required is usually much more of a problem.
4. Descriptions of real machine programs using simpler abstract models are often much more complex than descriptions using Turing machines. For example, a Turing machine describing an algorithm may have a few hundred states, while the equivalent deterministic finite automaton (DFA) on a given real machine has quadrillions. This makes the DFA representation infeasible to analyze.
5. Turing machines describe algorithms independent of how much memory they use. There is a limit to the memory possessed by any current machine, but this limit can rise arbitrarily in time. Turing machines allow us to make statements about algorithms which will (theoretically) hold forever, regardless of advances in *conventional* computing machine architecture.
6. Turing machines simplify the statement of algorithms. Algorithms running on Turing-equivalent abstract machines are usually more general than their counterparts running on real machines, because they have arbitrary-precision data types available and never have to deal with unexpected conditions (including, but not limited to, running out of memory).

An experimental prototype to achieve Turing machine

One way in which Turing machines are a poor model for programs is that many real programs, such as operating systems and word processors, are written to receive unbounded input over time, and therefore do not halt. Turing machines do not model such ongoing computation well (but can still model portions of it, such as individual procedures).

## Limitations of Turing Machines

### Computational Complexity Theory

A limitation of Turing machines is that they do not model the strengths of a particular arrangement well. For instance, modern stored-program computers are actually instances of a more specific form of abstract machine known as the random access stored program machine or RASP machine model. Like the Universal Turing machine the RASP stores its "program" in "memory" external to its finite-state machine's "instructions". Unlike the universal Turing machine, the RASP has an infinite number of distinguishable, numbered but unbounded "registers"—memory "cells" that can contain any integer (cf. Elgot and Robinson (1964), Hartmanis (1971), and in particular Cook-Rechow (1973); references at random access machine). The RASP's finite-state machine is equipped with the capability for indirect addressing (e.g. the contents of one register can be used as an address to specify another register); thus the RASP's "program" can address any register in the register-sequence. The upshot of this distinction is that there are computational optimizations that can be performed based on the memory indices, which are not possible in a general Turing machine; thus when Turing machines are used as the basis for bounding running times, a 'false lower bound' can be proven on certain algorithms' running times (due to the false simplifying assumption of a Turing machine). An example of this is binary search, an algorithm that can be shown to perform more quickly when using the RASP model of computation rather than the Turing machine model.

### Concurrency

Another limitation of Turing machines is that they do not model concurrency well. For example, there is a bound on the size of integer that can be computed by an always-halting nondeterministic Turing machine starting on a blank tape. By contrast, there are always-halting concurrent systems with no inputs that can compute an integer of unbounded size. (A process can be created with local storage that is initialized with a count of 0 that concurrently sends itself both a stop and a go message. When it receives a go message, it increments its count by 1 and sends itself a go message. When it receives a stop message, it stops with an unbounded number in its local storage.)

## History

They were described in 1936 by Alan Turing.

### Historical Background: Computational Machinery

Robin Gandy (1919–1995)—a student of Alan Turing (1912–1954) and his lifelong friend—traces the lineage of the notion of "calculating machine" back to Babbage (circa 1834) and actually proposes "Babbage's Thesis":

*That the whole of development and operations of analysis are now capable of being executed by machinery.*

*—(italics in Babbage as cited by Gandy, p. 54)*

Gandy's analysis of Babbage's Analytical Engine describes the following five operations (cf. p. 52–53):

1. The arithmetic functions +, −, × where − indicates "proper" subtraction x − y = 0 if y ≥ x
2. Any sequence of operations is an operation
3. Iteration of an operation (repeating n times an operation P)
4. Conditional iteration (repeating n times an operation P conditional on the "success" of test T)
5. Conditional transfer (i.e. conditional "goto").

Gandy states that "the functions which can be calculated by (1), (2), and (4) are precisely those which are Turing computable." . He cites other proposals for "universal calculating machines" including those of Percy Ludgate (1909), Leonardo Torres y Quevedo (1914), Maurice d'Ocagne (1922), Louis Couffignal (1933), Vannevar Bush (1936), Howard Aiken (1937). However:

... the emphasis is on programming a fixed iterable sequence of arithmetical operations. The fundamental importance of conditional iteration and conditional transfer for a general theory of calculating machines is not recognized ...

*—Gandy p. 55*

The Entscheidungsproblem (the "decision problem"): Hilbert's tenth question of 1900

With regard to Hilbert's problems posed by the famous mathematician David Hilbert in 1900, an aspect of problem #10 had been floating about for almost 30 years before it was framed precisely. Hilbert's original expression for #10 is as follows:

10. Determination of the solvability of a Diophantine equation. Given a Diophantine equation with any number of unknown quantities and with rational integral coefficients: To devise a process according to which it can be determined in a finite number of operations whether the equation is solvable in rational integers. The Entscheidungsproblem [decision problem for first-order logic] is solved when we know a procedure that allows for any given logical expression to decide by finitely many operations its validity or satisfiability ... The Entscheidungsproblem must be considered the main problem of mathematical logic.

*—quoted, with this translation and the original German, in Dershowitz and Gurevich, 2008*

By 1922, this notion of "Entscheidungsproblem" had developed a bit, and H. Behmann stated that

... most general form of the Entscheidungsproblem [is] as follows:

> A quite definite generally applicable prescription is required which will allow one to decide in a finite number of steps the truth or falsity of a given purely logical assertion ...

*—Gandy p. 57, quoting Behmann*

Behmann remarks that ... the general problem is equivalent to the problem of deciding which mathematical propositions are true.

*—ibid.*

If one were able to solve the Entscheidungsproblem then one would have a "procedure for solving many (or even all) mathematical problems".

*—ibid., p. 92*

By the 1928 international congress of mathematicians, Hilbert "made his questions quite precise. First, was mathematics *complete* ... Second, was mathematics *consistent* ... And thirdly, was mathematics *decidable*?" (Hodges p. 91, Hawking p. 1121). The first two questions were answered in 1930 by Kurt Gödel at the very same meeting where Hilbert delivered his retirement speech (much to the chagrin of Hilbert); the third—the Entscheidungsproblem—had to wait until the mid-1930s.

The problem was that an answer first required a precise definition of "*definite general applicable prescription*", which Princeton professor Alonzo Church would come to call "effective calculability", and in 1928 no such definition existed. But over the next 6–7 years Emil Post developed his definition of a worker moving from room to room writing and erasing marks per a list of instructions (Post 1936), as did Church and his two students Stephen Kleene and J. B. Rosser by use of Church's lambda-calculus and Gödel's recursion theory (1934). Church's paper (published 15 April 1936) showed that the Entscheidungsproblem was indeed "undecidable" and beat Turing to the punch by almost a year (Turing's paper submitted 28 May 1936, published January 1937). In the meantime, Emil Post submitted a brief paper in the fall of 1936, so Turing at least had priority over Post. While Church refereed Turing's paper, Turing had time to study Church's paper and add an Appendix where he sketched a proof that Church's lambda-calculus and his machines would compute the same functions.

But what Church had done was something rather different, and in a certain sense weaker. ... the Turing construction was more direct, and provided an argument from first principles, closing the gap in Church's demonstration.

*—Hodges p. 112*

And Post had only proposed a definition of calculability and criticized Church's "definition", but had proved nothing.

## Alan Turing's a-machine

In the spring of 1935, Turing as a young Master's student at King's College Cambridge, UK, took on the challenge; he had been stimulated by the section of the logician M. H. A. Newman "and learned from them of Gödel's work and the Entscheidungsproblem ... Newman used the word 'mechanical' ... In his obituary of Turing 1955 Newman writes:

To the question 'what is a "mechanical" process?' Turing returned the characteristic answer 'Something that can be done by a machine' and he embarked on the highly congenial task of analysing the general notion of a computing machine.

—*Gandy, p. 74*

Gandy states that:

I suppose, but do not know, that Turing, right from the start of his work, had as his goal a proof of the undecidability of the Entscheidungsproblem. He told me that the 'main idea' of the paper came to him when he was lying in Grantchester meadows in the summer of 1935. The 'main idea' might have either been his analysis of computation or his realization that there was a universal machine, and so a diagonal argument to prove unsolvability.

—*ibid., p. 76*

While Gandy believed that Newman's statement above is "misleading", this opinion is not shared by all. Turing had a lifelong interest in machines: "Alan had dreamt of inventing typewriters as a boy; [his mother] Mrs. Turing had a typewriter; and he could well have begun by asking himself what was meant by calling a typewriter 'mechanical'" (Hodges p. 96). While at Princeton pursuing his PhD, Turing built a Boolean-logic multiplier. His PhD thesis, titled "Systems of Logic Based on Ordinals", contains the following definition of "a computable function":

It was stated above that 'a function is effectively calculable if its values can be found by some purely mechanical process'. We may take this statement literally, understanding by a purely mechanical process one which could be carried out by a machine. It is possible to give a mathematical description, in a certain normal form, of the structures of these machines. The development of these ideas leads to the author's definition of a computable function, and to an identification of computability with effective calculability. It is not difficult, though somewhat laborious, to prove that these three definitions [the 3rd is the λ-calculus] are equivalent.

—*Turing (1939) in The Undecidable, p. 160*

When Turing returned to the UK he ultimately became jointly responsible for breaking the German secret codes created by encryption machines called "The Enigma"; he also became involved in the design of the ACE (Automatic Computing Engine), "[Turing's] ACE proposal was effectively self-contained, and its roots lay not in the EDVAC [the USA's initiative], but in his own universal machine" (Hodges p. 318). Arguments still continue concerning the origin and nature of what has been named by Kleene (1952) Turing's Thesis. But what Turing *did prove* with his computational-machine model appears in his paper *On Computable Numbers, With an Application to the Entscheidungsproblem* (1937):

[that] the Hilbert Entscheidungsproblem can have no solution ... I propose, therefore to show that there can be no general process for determining whether a given formula U of the functional calculus K is provable, i.e. that there can be no machine which, supplied with any one U of these formulae, will eventually say whether U is provable.

*—from Turing's paper as reprinted in The Undecidable, p. 145*

Turing's example (his second proof): If one is to ask for a general procedure to tell us: "Does this machine ever print 0", the question is "undecidable".

## 1937–1970: The "Digital Computer", the Birth of "Computer Science"

In 1937, while at Princeton working on his PhD thesis, Turing built a digital (Boolean-logic) multiplier from scratch, making his own electromechanical relays (Hodges p. 138). "Alan's task was to embody the logical design of a Turing machine in a network of relay-operated switches ..." (Hodges p. 138). While Turing might have been just initially curious and experimenting, quite-earnest work in the same direction was going in Germany (Konrad Zuse (1938)), and in the United States (Howard Aiken) and George Stibitz (1937); the fruits of their labors were used by the Axis and Allied military in World War II (cf. Hodges p. 298–299). In the early to mid-1950s Hao Wang and Marvin Minsky reduced the Turing machine to a simpler form (a precursor to the Post-Turing machine of Martin Davis); simultaneously European researchers were reducing the new-fangled electronic computer to a computer-like theoretical object equivalent to what was now being called a "Turing machine". In the late 1950s and early 1960s, the coincidentally parallel developments of Melzak and Lambek (1961), Minsky (1961), and Shepherdson and Sturgis (1961) carried the European work further and reduced the Turing machine to a more friendly, computer-like abstract model called the counter machine; Elgot and Robinson (1964), Hartmanis (1971), Cook and Reckhow (1973) carried this work even further with the register machine and random access machine models—but basically all are just multi-tape Turing machines with an arithmetic-like instruction set.

## 1970–present: the Turing Machine as a Model of Computation

Today, the counter, register and random-access machines and their sire the Turing machine continue to be the models of choice for theorists investigating questions in the theory of computation. In particular, computational complexity theory makes use of the Turing machine:

Depending on the objects one likes to manipulate in the computations (numbers like nonnegative integers or alphanumeric strings), two models have obtained a dominant position in machine-based complexity theory:

> *the off-line multitape Turing machine*..., which represents the standard model for string-oriented computation, and
>
> the *random access machine (RAM)* as introduced by Cook and Reckhow ..., which models the idealized Von Neumann style computer.

*—van Emde Boas 1990:4*

Only in the related area of analysis of algorithms this role is taken over by the RAM model.

*—van Emde Boas 1990:16*

## Thinking Humanly: The Cognitive Modeling Approach

The interdisciplinary field of cognitive science brings together computer models from Artificial Intelligence and experimental techniques from cognitive psychology to try to construct precise and testable theories of the workings of the human mind. And if we are going to say that a given program thinks like a human being, we must have some way of determining how human beings think. For that, we need to get inside the actual workings of the human mind. Stuart Russell and Peter Norvig say, "there are two ways to do this: through introspection–trying to catch our own thoughts as they go by –or through psychological experiments. Once we have a sufficiently precise theory of the mind, it becomes possible to express the theory as a computer program. If the program's input/output and timing behavior matches human behaviour, that is evidence that some of the program's mechanisms may also be operating in humans."

Now it is almost taken for granted by many psychologists that a cognitive theory should be like a computer program. But we know that cognitive science is the science of mind. Therefore cognitive scientists seek to understand perceiving, thinking, remembering, understanding language, learning and other mental phenomenon. Their research is remarkably diverse, ranging from observing children's mental operation, through programming computers to do complex problem solving, to analyze the nature of meaning. In order to appreciate the work in artificial intelligence, which is a necessary part of cognitive science, it is necessary to have some familiarity with theories of human intelligence. The cognitive scientists introduce the notion of machine intelligence and emphasize the relationship between human and machine intelligence. The aim of artificial intelligence is to develop and test computer programs that exhibit characteristic of human intelligence. The most fundamental contribution of symbolic computational modeling has been the physical symbol system hypothesis. According to Newell and Simon, "a physical symbol system has the necessary and sufficient means for general intelligent action," By 'necessary' we mean that any system that exhibits general intelligence will prove upon analysis to be a physical symbol system. By 'sufficient' we mean that any physical symbol system of sufficient size can be organized further to exhibit general intelligence. Lastly, by 'general intelligent action' we wish to indicate the same scope of intelligence as we see in human action; that in any real situation behaviour appropriate to the events of the system and adaptive to the demands of the environment can occur within some limits.

However, the ability of computer simulations to model such process is interpreted as a proof of the broader claim that a symbol system is at the center of human intelligence. In this hypothesis it shows that intelligence is an essential aspect of machines. If the machines have the capacity of intelligence, intelligence is the essence of human cognitions. Therefore, machines have cognitive capacity like the human beings. In the cognitive modeling approach, thus, human beings and machines show the property of being intelligent.

## Thinking Rationally: The Laws of Thought Approach

"Right thinking" is the inferential character of every reasoning process. Aristotle in his famous syllogisms provided patterns of argument structures that always give correct conclusions from given correct premises. In the syllogisms, the Laws of Thought play a vital role because these give law the right explanation of a syllogistic inference. There are three Laws of Thought recognized by the logicians. These have traditionally been called the law of Identity, the law of Contradiction, and the law of Excluded–Middle.

These Laws of Thought are appropriate to different contexts. The formulations appropriate as follows:

a. The law of Identity asserts that *if any statement is true, then it is true*. This law asserts that every statement of the form $P \supset P$ is true, and that every such statement is a tautology.

b. The law of Contradiction asserts that *no statement can be both true and false*. This law asserts that every statement of the form $P \,.\, {\sim}P$ is false, that is, every such statement is self-contradictory, and its negation is logically true.

c. The law of Excluded-Middle asserts that *any statement is either true or false*. This law asserts that every statement of the form $P \vee {\sim}P$ is true, that is, every such statement is a tautology

In the 'Laws of Thought' approach to artificial intelligence, the whole emphasis is on correct syllogistic inferences. For example-

"Socrates is a man;

All men are mortal;

Therefore, Socrates is mortal."

In this inference, the conclusion is based on the premises according to the rules of inference. The above syllogistic inference is the best example to formulate an AI program. In all reasoning of this type the emphasis is on the logical inference of a conclusion from the premises. In the AI programme this type of logical inference is of much use since this programme provides a variety of logical reasoning. In an inference, a set of variables, a set of constant terms, a set of functions, the set of connectives *if, and, or,* and, *not,* the quantifiers *'exists'* and *'for all'* are the most important symbols to build an AI program. All these constants and variables are the arbitrary representations of the world. With the help of these symbols, the so-called logistic tradition within artificial intelligence hopes to build on such programs to create intelligent systems.

## Acting Rationally: The Rational Agent Approach

Here, the word 'agent' refers to mechanical agent (computer). Acting rationally means acting so as to achieve one's goal, given one's beliefs. An agent (mechanical) is something that perceives and acts. Stuart Russell and Peter Norvig point out that making correct inferences is a part of being a rational agent because one way to act rationally is to reason logically to the conclusion that a given action will achieve one's goals, and then to act on that conclusion.

According to them, "the study of artificial intelligence as rational agent design therefore has two advantages. First, it is more general than the 'laws of thought' approach, because correct inference is only a useful mechanism for achieving rationality, and not a necessary one. Second, it is more amenable to the scientific development than approaches based on human behavior or human thought, because the standard of rationality is clearly defined and completely general." As we have seen, an agent is something that perceives and acts in an environment. The job of artificial intelligence is to design the agent program, that is a function that implements the agent mapping from percepts to action. We assume this program will run on some sort of computing device A human agent has eyes, ears, and other organs for sensors, and hands, legs, mouth and other body parts for effectors. The relationship among agents, architectures, and programs can be summed up as: *Agent = Architecture + Program.*

The agent is autonomous to the extent that its behaviour is determined by its own experience. A truly autonomous intelligent agent should be able to operate successfully in a wide variety of environments, given sufficient time and scope. But before we design an agent, we must have a pretty good idea of the possible percepts and actions, the agent's goal is supposed to achieve, and what sort of environment it will operate in it.

Again, as we have mentioned, an agent is just something that perceives and acts. In this approach, artificial intelligence is viewed as the study and construction of a rational agent. One of the important factors is that correct inference is not the whole of rationality. There are also ways of acting rationally that cannot reasonably be said to involve inference. For example, pulling one's hand off a hot stone is a reflex action that is more successful than a slower action taken after careful deliberation.

## The Field of Artificial Intelligence

Artificial intelligence is a new research area of growing interdisciplinary interest and practical importance. People with widely varying backgrounds and professional knowledge are contributing new ideas and introducing new tools into this discipline. Cognitive minded psychologists have developed new models of the mind based on the fundamental concepts of artificial intelligence, symbols systems and information processing. Linguists are also interested in these basic notions while developing different models in computational linguistics. And philosophers, in considering the progress, problems and potential of this work towards non-human intelligence, have sometimes found solution to the age-old problems of the nature of mind and knowledge.

However, we know that artificial intelligence is a part of computer science in which are designed intelligent systems that exhibit the characteristics we associate with intelligence in human behaviour, understanding language learning, reasoning, problem solving, and so on. It is believed that insights into the nature of the mind can be gained by studying the operation of such systems. Artificial intelligence researchers have invented dozens of programming techniques that support some sort of intelligent behaviour.

Now the question is: Is artificial intelligence a science or an Art? According to Tanimoto, "the activity of developing intelligent computer systems employs both proved mathematical principles, empirical results of studying previous system, and heuristic, pragmatic programming techniques. Information stored in rational data structures can be manipulated by well–studied techniques of computer science such as tree searching algorithms. At the same, experimental or vaguely understood 'rules of thumb' for problem solving are often crucial to the success of a system and must be carefully accommodated in intelligent systems." Thus artificial intelligence is both an art and a science. It is a science because it develops intelligent computer systems by employing proved mathematical principles. It is an art because it designs systems by employing programming techniques. Information stored in relational data structure can be manipulated by well-studied techniques of computer science such as tree searching diagram. Thus, the field of artificial intelligence is fascinating because of this complementing of art and science.

Artificial intelligence research may have impact on science and technology in the following way:

i. It can solve some difficult problems in chemistry, biology, geology, engineering and medicine.

ii. It can manipulate robotic devices to perform some useful, repetitive, sensory-motor tasks;

Besides, artificial intelligence researchers investigate different kinds of computation and different ways of describing computation in an effort not just to create intelligent artefacts, but also to understand what intelligence is. According to Charniak and McDermott, their basic tenet is to create computers which think. Thus AI expands the field of intelligent activity of human beings in various ways.

Now the question is: What would the world be like if we had intelligent machines? What would the existence of such machines say about the nature of human beings and their relation to the world around them? These questions have raised profound philosophical issues which will be discussed in due course.

The hypothesis of artificial intelligence and its corollaries are empirical in nature whose truth or falsity is to be determined by experiment and empirical test. The method of testing the results of artificial intelligence is of the following types:

i. In the narrow sense, artificial intelligence is part of computer science, aimed at exploring the range of tasks over which computers can be programmed to behave intelligently. Thus it is the study of the ways computers can be made to perform cognitive tasks, which generally human beings undertake.

ii. In the wider sense, artificial intelligence is aimed at programs that simulate the actual processes that human beings undergo in their intelligent behaviour. And these simulation programs are intended as theories describing and explaining human performance. But they are tested by comparing the computer output with the human behavior to determine whether both the result and also the actual behavior of computers and persons are closely similar.

A digital computer is an example of a physical symbol system, a system that is capable of inputting, outputting, storing, etc., following different courses of operation. These systems are capable of producing intelligence depending on the level of mechanical sophistication they are. The computers with these capabilities behave intelligently like human beings, according to the AI researchers.

One of the important tenets of cognitive science is that it stresses the relationship between human and machine intelligence. It is interested in using artificial intelligence techniques to enlighten us about how human beings do intelligent tasks. For example, in getting a machine to solve geometric or integral calculus problems, we are also interested in learning more about the power and flexibility of the human capacity for problem solving.

Thus, the AI researcher attempts to develop and test computer programs that exhibit characteristics of human intelligence. The AI researcher always works with an artifact, which is only inferentially related to the human mind. As Boden says, "The 'machines' in question are typically digital computers, but artificial intelligence is not the study of computers." But it is the study of intelligence in thought and action. Moreover, as Bonnet would say, "artificial intelligence is the discipline that aims to understand the nature of human intelligence through the construction of computer programs that imitate intelligent behaviour."

Further, Bonnet says, "we cannot define human intelligence in general we can highlight a number of criteria by which it can be judged, such as the ability to make abstractions or generalizations, to draw analogies between different situations and to adopt new ones, to detect and correct mistakes in order to improve future performance, and so on" AI is concerned with the intelligent properties of the computational systems, which perform many intelligent tasks. These tasks can be of very varied nature, such as, the understanding of a spoken or written text in a natural language, playing chess, solving a puzzle, writing a poem, making a medical diagnosis, finding one's way from Hyderabad to Delhi. The AI programme selects such activities include information and reasoning processes which are part of any intelligent system.

Artificial intelligence includes commonsensical tasks, such as understanding English language, recognizing scenes, finding a way to reach an object that is far overhead, heavy and frigid, and making sense of the plot of a mystery novel. In addition, artificial intelligence includes expert tasks, such as diagnosing diseases, designing computer systems, locating mineral deposits, and planning scientific experiments. The techniques that artificial intelligence applies to solve these problems are representation and inference methods for handling the relevant knowledge, and search based problem-solving methods for exploiting that knowledge. Although the tasks with which artificial intelligence is concerned may seem to form a very heterogeneous set, in fact, they are related through their common reliance on techniques for manipulation of knowledge and conducting search.

## What Computers Can Do

Within the scientific discipline of artificial intelligence, there are several distinct areas of research, each with its own specific interests, research techniques, and terminology. However, as Avron Barr and Edward A. Feigenbaum point out, artificial intelligence is part of computer science concerned with designing intelligent computer systems, that is, systems that exhibit the characteristics we associate with intelligence in human behavior such as understanding language, learning, reasoning, solving problems and so on.

In artificial intelligence, this specialization includes research on language understanding, automatic programming, and several others. The following discussion of the status of artificial intelligence attempts to show the sub–fields identifying some aspects of intelligent behaviours and indicating the state of relevant research.

Like the different sub–fields of artificial intelligence, the different behaviours discussed here are not at all independent; separating them out is just a convenient way of indicating what current artificial intelligence programs can do.

i. The Problem Solving

The first big successes in artificial intelligence were programs that could solve puzzles and play games like chess. Techniques like looking ahead several moves and dividing difficult problems into easier sub-problems evolved in the fundamental artificial intelligence techniques of search and problem reduction. Another problem solving program that integrates mathematical formulas symbolically has attained very high level of performance and is being used by scientists and engineers. According to Avron Barr and Edward A. Feigenbaum, human beings often solve a problem by finding a way of thinking about it that makes the solution easy, but so far, artificial intelligence programs must be told how to think about the problems they solve.

Elaine Rich defines problem solving, as that which "chooses the best technique(s) and applies it (them) to the particular problem." Another view is that a goal and a set of means of achieving the goal are called a problem solving. Before an agent can start searching for solutions, it must formulate a goal and then use the goal to formulate a problem. According to Stuart Russell and Peter Norvig, a problem consists of four parts. The initial state, a set of operators, a goal test function, and a path cost function. The environment of the problem is represented by a state space. A path through the state space from the initial state to a goal state is a solution. And a search can be judged on the basis of completeness, optimality, time complexity, and space complexity. There are different search paths for an intelligent agent. The agent has to search which is the minimum path to reach the goal. To solve a problem is the main task of the agent in artificial intelligence.

Moreover, one of the important searches in artificial intelligence is the technique of 'heuristic' search. The word 'heuristic' is derived form the Greek verb 'heuriskein', meaning 'to find' or 'to discover'. Some people use heuristic as the opposite of algorithmic. According to Newell, Shaw and Simon, "process that may solve a given problem, but offers no guarantees of doing so is called a heuristic for that problem."

We know that heuristic techniques dominated early application of artificial intelligence. Heuristic method is still used in problem solving. For example, game playing is also a form of problem solving. Games have engaged the intellectual faculties of the humans. And game playing is also one of the oldest areas of endeavour in artificial intelligence. A chess-playing computer is a proof of a machine doing something though intelligence. Furthermore, the simplicity of the rules and their application in the program implies that it is easy to represent the game as a search through a space of possible game positions.

The initial state, the operators, a terminal test, and a pay-off function can define a game. According to Carpenter and Just, an intelligent machine can follow all these rules and play more efficiently than human beings. For example, in speed chess, computers have defeated the world champion, Gary Kosparov, in both five-minutes and twenty five minutes games. Such a system would be a significant achievement not just for game playing research, but also for artificial intelligence research in general, because it would be much more likely to be applied to the problem faced by a general intelligent agent. There are so many kinds of search in order to solve a problem. But we are not explaining the entire search. We are mainly concerned with how artificial intelligence programs solve different problems within a few seconds.

## Natural Language Processing

The most common way that people communicate is by speaking or writing in one of the natural languages like, Hindi, English, French or Chinese. However, the computer-programming language seems to differ from human languages. These 'artificial' languages are designed so that sentences have a rigid format, or syntax, making it easier for compliers to phrase the programs and convert them into the proper sequences of computer instructions. Besides, being structurally simpler than natural language, programming language can easily express only those concepts that are important in programming: 'Do this, then do that', 'See whether such and such is true'.

One of the important facts is that if computers could understand what people mean when people use English language, the systems would be easier to use and would fit more naturally into people's lives. Artificial intelligence researchers hope that learning how to build computers that can communicate as people do would extend our understanding of human language and mind.

The goal of cognitive linguistics is to specify a theory of language comprehension and production to such a level of detail that a person could write a computer program that could understand and produce natural language. Thus, the intelligent machine program has the capacity to understand natural language. It also responds to questions in an appropriate manner. Thus, if the question were 'Is Rome the capital of France?' the simple reply 'No' would be appropriate but, 'no, it is Paris,' or, 'No, Rome is the capital of Italy' would be more complex. Machine programs would also be able to explain the meaning of the word in other terms and also to translate from one language to another.

## Machine Vision

Vision is the information processing of understanding a scene from its projected images. An image is a two-dimensional function (X, Y) obtained, with a sensing device that records the value of an image feature at all points (X, Y). Images are converted into a digital form for processing with a computer. The task of a computer-vision system is to understand the scene that an imaginary of pixels depicts.

Generally, machine vision means the vision of a machine that can see or perceive the things in the world. According to David Marr, vision is the process of discovering form images what is present in the world, and where it is. Thus, vision is, first and foremost, an information-processing task. Moreover, there is a distinction between human vision (perception) and machine vision, because human vision is defined as "the eye is just a sensor; the visual cortex of the human brain is our primary organ of vision." In the case of human vision, the sensory organ necessarily forms a representation or physical encoding of the received information that facilitates the answering of some questions about the environment, but makes it extremely difficult to answer others. For example, an examination of the encoded information produced by the human eye reveals that at any instant of time the eye has sensed only a small part of the electromagnetic spectrum, and has extracted an image of the scene from a particular viewpoint in space. It possesses the visual information with the help of the lens fitted within it. Thus machine vision works more or less like human vision.

Learning is now perceived as a gateway to understanding the problem of intelligence. Because seeing is also intelligence, seeing is also learning which is a key to the study of intelligence and biological vision. Visual neuroscience develops the methods of understanding how human visual systems work. Visual neuroscience is beginning to focus on the mechanisms that allow the cortex to adapt its circuitry and learn a new task. Machine vision as a part of the AI programme, however, tries to develop systems that can simulate human vision.

## Machine Learning

Learning for an intelligent agent is essential for dealing with unfamiliar environments. Learning a function from examples of its input and output is called inductive reasoning. According to Bonnet, the ability to learn is one of the fundamental constituents of intelligence, 'learning' being understood in its general sense as indicating the way in which the humans and animals can increase their stock of knowledge and improve their skill and reasoning powers.

Moreover, from the very beginning of artificial intelligence, researchers have sought to understand the process of learning and to create computer programs that show learning behaviour. Ordinarily, learning is a very general term denoting the way in which people increase their knowledge and improve their skills. There are two fundamental reasons for studying learning. One is to understand the process itself. By developing computer models of learning, psychologists have attempted to gain an understanding of the way humans learn. Philosophers since Plato have also been interested in the process of learning, because it might help them understand and know what knowledge is and how it grows.

The second reason for learning research is to provide computers with the ability to learn. Learning research has potential for extending the range of problems to which computers can be applied. In particular, the work in machine learning is important for expert systems development, problem solving, computer vision, speech understating, conceptual analysis of databases, and intelligent authoring systems.

In this way, a computer can do more work than human beings. According to James R. Slagle, computers can do arithmetic at an unbelievable speed, and no human can compete with them. Besides, computers can do many other odd jobs like detecting a certain disease from the available data. Thus, AI as a study of machine intelligence, especially of computers and robots, opens up new vistas of understanding mind and intelligence. Intelligent behaviour can be studied in many dimensions. This is made possible by the invention of computers as high-processing computing machines. The computers can do many things that seem to require intelligence and other mental abilities.

## The Computational Model of Mind

As we have already seen, artificial intelligence is the discipline that aims to understand the nature of human intelligence through the construction of computer programs that imitate intelligent behavior. It also emphasizes the functions of the human brain and the analogical functioning of the digital computer. According to one extreme view, the human brain is just a digital computer and the mind is a computer program. This view, as John Searle calls it is strong artificial intelligence.

According to strong artificial intelligence, "the appropriately programmed computer with the right inputs and outputs literally has a mind in exactly the same sense that you and I do." This tells that not only the devices would just referred to indeed be intelligent and have minds, etc. but mental qualities of a sort can be attributed to teleological functioning of any computational device, even the very simplest mechanical ones such as a thermostat. Here, the idea is that mental activity is simply the carrying out of some well-defined operations, frequently referred as an algorithm. We may ask here as to what an algorithm actually is. It will be adequate to define an algorithm simply as a calculation procedure of some kind. But in the case of thermostat, the algorithm is extremely simple: the device registers whether the temperature is greater or smaller than the setting, and then, it arranges for the circuit to be disconnected in the former case and to remain connected in the latter. For understanding any significant kind of mental activity of a human brain, a very complex set of algorithms has to be designed to capture the complexity of the human mental activities. The digital computers are approximations to the complex human brain.

The strong artificial intelligence view is that the differences between the essential functioning of a human being (including all its conscious manifestations) and that of a computer lies only in the much greater complication in the case of brain. All mental qualities such as thinking, feeling, intelligence, etc. are to be regarded, according to this view, merely as aspects of this complicated functioning of the brain; that is to say that they are the features of the algorithm being carried out by the brain. The brain functions like a digital computer according to this view.

The supporters of strong AI hold that the human brain functions like a Turing machine which carries out all sets of complicated computations. The brain is naturally designed like a computing machine to think, calculate and carry out algorithmic activities. To strong AI supporters, the activities of the brain are simply algorithmic activities which give rise to all mental phenomena like thinking, feeling, willing, etc.

David Chalmers, mentions that the field of artificial intelligence is devoted in large part to the goal of reproducing mentality in computational machines. The supporters of strong AI argue that we have every reason to believe that eventually computers will truly have minds. Winston says that, "intelligent robots must sense, move and reason." Accordingly, intelligent behaviour is interpreted as giving rise to abstract automation. That is to say that an artificial, non-biological system could thus be the sort of thing that could give rise to conscious experience. For the supporters of strong AI, humans are indeed machines, and in particular, our mental behaviour is finally the result of the mechanical activities of the brain.

John Searle, in his "*Is the Brain a digital computer*?" mentions that the basic idea of the computer model of the mind is that the mind is the software and the brain is the hardware of a computational system. The slogan is: "the mind is to the program, the brain is to the hardware." For strong AI, there is no distinction between brain processes and mental processes. Because the process which is a happening in the brain is a computational process, the mind is the alternative name of the brain which is a machine.

According to David Chalmers, the theory of computation deals wholly with abstract objects such as Turing machine, Pascal program, finite state automation and so on. These abstract objects are formal structures which are implemented in formal systems. However, the notion of implementation is the relation between abstract computational objects and physical systems. Computations are often implemented in synthetic silicon based computers.

Whereas the computational systems are abstract objects with a formal structure determined by their states and state transition relations, the physical systems are concrete objects with a causal structure determined by their internal states and the causal relations between the states. It may be pointed out that a physical system implements a computation when the casual structure of the system mirrors the formal structure of the computation. The system implements the computation, if there is a way of mapping states of the system onto states of the computations so that the physical states which are causally related map onto the formal stares that are correspondingly formally related.

The fact is that there is rich causal dynamics inside computers, as there is in the brain. There is real causation going on between various units of brain activity precisely mirroring patterns of causation between the neurons. For each neuron, there is a specific causal link with other neurons. It is the causal patterns among the neurons in the brain that are responsible for any conscious experiences that may arise.

The brain, as Marvin Minsky says, "happens to be a meat machine." He points out that the brain is an electrical and chemical mechanism, whose organization is enormously complex, whose evaluation is barely understood and which produces complex behavior in response to even more complex environment. Artificial intelligence understands the nature of human intelligence in terms of the computational model of the brain.

## Artificial Intelligence and the Functionalist Theory of Mind

Functionalism arose as a result of the phenomenal rise of interest in computing machines and artificial intelligence. The functionalists say that mental processes are computational processes realized in a machine. Functionalism is a theory that explains mental phenomenona in terms of the external input and the observable output. It explains the mind as a machine.

Functionalism as a conception of mind explains the reality of the mental phenomena in terms of the mechanical phenomena. Mind is not just a part of the physical organism but is itself characterized as the physical organism called the brain. The brain is further characterized as a complex physical system working like a computing machine. Functionalism explains the mechanical behaviour of the brain/mind.

Functionalism as a theory of mind is supported by various scientific theories like those of artificial intelligence, cognitive science, and neuroscience, etc., Artificial intelligence advocates a computational theory of mind which argues in favour of the functional similarity between the computation states of the artificial system and the neurophysiological states of the brain.

The hypothesis of artificial intelligence that "machine can think" became very popular after Alan Turing's "*Computing Machine and Intelligence*." Turing's hypothesis is that machines think intelligently like human beings. Putnam writes, "Probabilistic automation has been generalized to allow for 'sensory inputs', and 'motor outputs', -that is, the Machine Table specifies, for every possible combination of 'state' and a complete state of 'sensory inputs' an 'instruction' which determines the probability of the next 'state' and also probabilities of the 'motor out puts'." The following are the steps which explain how a machine functions in general:

i. The description of the sequence of states (procedure)

ii. Description of rules

iii. The explanation of the rationale of the entire procedure.

The computing machine is thus a system constructed out of different subsystems that function inside it to process the inputs and to produce the output once the input is simulated in the machine. It tries to match the simulating state with the states already computed and mapped in the system. This mapping order follows certain syntax or rules. The syntax is responsible for the correlation of total cognitive states. Thus, the entire process of simulation can be called an intelligent process. This simulation process takes place between the functions of the two functionally isomorphic systems. As Putnam defines, "Two systems are functionally isomorphic if there is a correspondence between states of one and the states of the other that preserves the functional relation."

There is functional isomorphism, according to Putnam, between the brain/mind and a machine. This functional isomorphism holds due to the causal capacity of the functional states of the machine. For example, when I have a pain, there is a neurophysiological process corresponding to the mental state because of the firing of the C-fiber. The brain/mind identity follows as there is functional identity between the two. Thus, identity between the mental states and the physical processes of the brain is established from the functional point of view. That is, in functional terms, the brain state is isomorphic with the mental state. That is, there is identity between software that constitutes the program and the hardware of the machine, which helps the software to be realized in the machine.

There can be indefinitely many different physical properties, which constitute the realizations of the same functional property. However, it is also true that the same physical state can realize different functional properties at different times or in different circumstances or in different creatures. The functional states are 'multiply realizable' in the sense that a functional state cannot be identical to any particular physical realization of it. For example, someone could write a program using two completely different types of computer, which use different sorts of hardware to run the same program. In this sense, the program is said to be 'multiply realizable' in that any number of computers may be used to realize the same program. Functionalism takes states of mind and mental properties to be functional states and properties. Mental properties are realizable by, but not identical with, material properties. For example, the same mental property, the property of being in pain, may be realized by one property in a human being and to a certain extent by another property in an invertebrate. For the functionalist, if someone has now a particular pain, then he/she can imagine that this pain is realized through a particular neural state. That neural state has an identifiable material structure, and this may be studied by a lower-level hardware science like neurobiology. Therefore, for functionalism, what makes the state a realization of pain, is not its material constitution but it's occupying a particular kind of causal role within our nervous system. Multiple realizability thus implies that there is a higher-level functional description of physical states in terms of their causal role, which abstracts from their lower-level physical constitution. It is with such functional properties that mental properties can be identified.

In his essay "*Mad Pain and Martian Pain*", David Lewis discusses two kinds of beings, which experience pain differently than normal humans. In the case of mad pain, the subject experiences pain when doing moderate exercise in an empty stomach; further, it improves his concentration for mathematical reasoning. On the other hand, Martian pain takes place in a Martian organism constructed of hydraulic hardware rather than neurons. Here the point is that pain is associated only contingently with either its causes (as in mad pain) or its physical realization (as in Martian pain). We cannot specify *a priori* its causal role or physical

Daniel Dennett has suggested a multiple-draft-model, approach to the nature of mind. According to this model, there is similarity between the functions of the human mind and those of the computer. The brain system functions in relation to different sub-systems. So there are multiple drafts, which operate within an artificial system. Such an analogy is beneficial because it analyses consciousness from the point of view of language processing. This is given importance precisely in the sense that a linguistic or language speaking being is considered not only as a conscious being but also a rational being. Even the robots as information processing systems can also be characterized as intelligent systems. According Dennett, "of course, we are machines! we are just very, very complicated, evolved machines made of organizes molecules instead of metal and silicon, and we are conscious, so there can be conscious machines – us."So the human thought process and language processing in the artificial systems are analogous to each other. In the case of the conscious thought process, we are aware of our thoughts, at the same time, there is physico-chemical process, which goes on in our brain.

Dennett's functional analysis of consciousness is divided into two parts. There are the sub-personal view of consciousness and the multiple draft-model of consciousness respectively. The sub-personal model explains consciousness and other mental activities through the help of neurological states and processes of the organism, whereas the multiple-draft-model discusses how an artificial system behaves intelligently. Thus Dennett provides a functional explanation of consciousness at the sub-personal level. According to him, "Sub-personal theories proceed by analyzing a person into an organization of subsystems (organs, routines, nerves, faculties, components-even atoms) and attempting to explain the behaviour of the whole person as the outcome of the interaction of these subsystems. Thus in the present instance the short-coming emerged because the two access notions introduced computational access *simpliciter* and the computational access of a print-out faculty, were defined at the sub-personal level; if introduced into a psychological theory they would characterize relations not between a person and a body, or a person and a state of affairs or a person and anything at all, but rather, at best relations between parts of person (or there bodies) and other things."

Therefore, the sub-personal level of explanation of consciousness tries to explain not only how the human beings are systems of organism but also how the system is being constituted and how the various functions involved in different physiological parts of the organism function together. And that functional structure would help us in defining the capacity involved in causing consciousness or what we call conscious behaviour. A state of consciousness is simply one which exhibits a certain characteristic pattern of causal relations to other states, both mental and physical.

We know that human beings perform various activities; they learn language; acquire various knowledge states or belief states, and there are changes in their belief states, and so on. All these activities are independent biological activities of human life. Dennett anticipates that there would be a system whose programs would be such that it would be self-dependent in all its functions. That would be able to replace or stand parallel to human intelligence. Further, the functions of the artificial system help us in explaining the various mysterious features that are ascribed the human life. Thus, the strong notion of functionalism advocates the identity between the mental states and the brain processes. It also explains the different basic features of human beings such as consciousness, intentionality and subjectivity, etc, by bringing the feature of the functional isomorphism into account.

Functionalism holds that the mental states are abstract, functional states, characterized solely in terms of their casual relationships to each other, to input, and to outoput. Human purposive behaviour is then explained in terms of how this hypothesized system of states take the organism from sensory input to behaviour. Because functionalism insists upon a network of mental states, it insists upon the holism of the mental upon the way in which mental states operate together to explain behaviour. It accepts structure of mental states in which each is necessarily connected with the other. The mental states do not function in isolation; rather they function within the causal co-relationship with other mental states. The function of mental states also takes into account the effect of the environment in which the subject or agent is placed with the system that must be well equipped to receive the input from the environment and to produce the output.

The functionalist program has been strongly influenced by analogies drawn from computer science and AI, both in its general outlook and in several of its specific applications to problems about the nature of mind. Because a functional state is like a computational state of a computer. A computer program can be described as a functional organization of the hardware. As already discussed, the functionalists argue that mental states are like the 'information processing' states of a computer. Accordingly, for computer functionalism or artificial intelligence, the brain is a computer, and the mind is a computer program implemented in the brain. Thus artificial intelligence is strongly founded on a functionalist conception of mind. It is dependent on the idea that human functions like a digital computer with multifunction computational abilities.

## Machine Intelligence, Competence, and Creativity

As already discussed, machine intelligence or artificial intelligence is a physical symbol system, which has the necessary and sufficient means for general action. The above statement shows that machine intelligence has the capacity of competence and creativity. However, the system that exhibits general intelligence will prove upon analysis to be a physical symbol system. Further the physical symbol system of sufficient size can be organized further to exhibit general intelligence. The general intelligent action indicates the same scope of intelligence as we have seen in human action. The symbol system hypothesis implies that intelligence will be realized by a universal computer. It also asserts that the intelligent machine is a symbol system. When machines acquired enough memory to make it practicable to locate actual symbol system with the help of programs produce intelligent behaviour and solve many problems.

Moreover, many machine intelligence scientists use computers as tools, to help them create things they could not have created. For example, many scientists use sounds which no orchestra could produce. A visual artist may get ideas from computer graphics. We are concerned with those programs which produce aesthetically interesting creations. There are a number of programs which explore artistically interesting spaces, and a few which produce aesthetically acceptable results.

Boden has given the example of a famous painter, Harold Cohen. Cohen has written a series of programs which produce pleasing, and unpredictable line drawings. Cohen's programs explore a certain style of line drawing and a certain subject matter. As human artists have to know about the things they are depicting, so each of Cohen's programs needs an internal model of its subject matter. This model is not a physical object. It is a set of abstract rules, which specify not only the anatomy of the human body, but also how the various body-parts appear from various points of view.

AI or Machine intelligence programs that write stories are woefully inadequate compared with human storyteller. But the best of them get what strength they possess from their internal models of very general aspects of motivation. For example, a program is asked to write a story with the moral 'Never trust flatterers.' In this story, there are two characters, the Fox and Crow. The story as follows:

"Once upon a time, there was a dishonest Fox, named, Henry who lived in a cave, and a vain and trusting Crow named Joe who lived in an elm-tree. Joe had gotten a piece of cheese and was holding it in his mouth. One day, Henry walked from his cave, across the meadow to the elm-tree. He saw Joe Crow and the cheese and became hungry. He decide that he might get the cheese if Joe Crow spoke, so he told Joe that he liked his singing very much and wanted to hear him sing. Joe was very pleased with Henry and began to sing. The cheese fell out of his mouth, down to the ground. Henry picked up the cheese and told Joe Crow that he was stupid. Joe was angry, and did not trust Henry any more. Henry returned to his cave. The end."

This program can construct hierarchical plans, ascribing them to the individual characters; according to the sorts of motivation one would expect them to have. It can give one character a role in another's plan. But these roles need not be allocated randomly, but can depend on background interpersonal relations. And it can represent different sorts of communication between the characters, which constrain what follows in different ways. All these matters are represented as abstract schemata, which are used to produce the story-structure. Thus the above programs shows that machines have the capacity to produce creativity. Now the question may be raised as; whether programs have scientific discovery? According to Langley, programs designed can find a simple mathematical and classificatory relation as 'rediscovered' with the help of physical and chemical laws.

NET talk is another famous and important example of machine creativity and competence. As we know, this model's goal is successfully to negotiate the problem domain of text-to-speech transformation. NET talk took seven letter segments of text as input and mapped the target window of that input onto an output which coded for phonemes. A connectionists system is not trained but programmed. Let us raise a question: What does NET talk know? According to Clark, "Net talk does not in any sense understand what it is saying. But this is not the point. Likewise, I might learn roughly how to pronounce Chinese sequences without understanding them. Nonetheless, NET talk has gone from babble to an output which is lawfully disciplined with respect to its input. That strongly suggests that it has learnt something. The question, what?"

According to Lenat, automatic mathematician is a system that does not produce proofs, nor solve mathematical problems. Rather, it generates and explains mathematical ideas. Artificial machine starts with 100 very primitive mathematical concepts drawn from set theory, including sets, lists, equality and operations. These concepts are so basic that they do not even include the ideas of eliminating arithmetic. To begin with, the program does not know what an integer is, still less addition, subtraction, multiplication, and division.

Artificial machine hunches, like human hunches, are sometimes wrong. Nevertheless, it has come up with some extremely powerful notions. It produced many arithmetical concepts, including integer, square root, addition, and multiplication. It generates the fundamental theorem of number theory, though did not prove. And it suggested the interesting idea that every even number greater than two is the sum of two different primes. It has originated one major theorem which no one had ever thought of before. Here, artificial machine appears to be P-creative. According to Boden, there are two senses of creativity. One is psychological creativity (P-creativity) and the other is historical creativity (H-creativity). An idea is P-creative if the person in whose mind it arises could not have had it before; it does not matter how many times other people have already had the same idea. On the other hand, an idea is H-creative if it is P-creative and no one else has ever had it before. Thus artificial machine's creative is P-creative.

Hofstadter presented another important example. He designed a copycat; in this, a program can generate many different analogies, where contextually appropriate comparisons are favoured over inappropriate ones. It does not rely on ready-made, fixed, representations, but constructs its own representations in a context-sensitive way. Its new analogies and new perceptions develop together.

As we have seen, cognitive science tries to provide computational models of the mind, that is, computational simulations of human cognitive process. If creativity is not a computational process, it might still be possible to simulate it computationally, just as it is possible to simulate digestive process without the simulation itself being a digestive process. It might be possible to have machine models of human creative processes, even if machines themselves cannot be creative. As Dartnall said, "if machines cannot be creative then I doubt there is any significant sense in which they can be intelligent, for they will never 'have minds of their own.' I do mean this in the weak sense that they will always slavishly do what we tell them, but in a strong-sense, they will never be able to generate their own ideas. And I take it as axiomatic that if they cannot generate their own ideas they cannot be intelligent."

## Mind in Artificial Intelligence

This section will explore the states of mind in artificial intelligence. As we know the main aim of artificial intelligence is to reproduce mentality in machines. That is to say that AI aims at producing machines with mind. If we say that machines have minds, then we have to ascribe certain 'belief', 'knowledge', 'free will', 'intention', 'observations', etc. to a machine. In that case, the machines will perform intelligent tasks and thus will behave like human beings.

We may raise a question: Why should we want ascribe mental qualities to machines at all? According to MacCarthy, there are many reasons for ascribing belief and other mental qualities:

i. Although we may know the program, its state at a given moment is usually not directly observable, and we can explain performance of a machine only by ascribing beliefs and goals to it.

ii. Ascribing beliefs may allow the derivation of general statements about the machine's behaviour that could not be obtained from any finite number of simulations.

iii. The difference between this program and another actual or hypothetical program may best be expressed as a difference in belief structure.

According to Haugeland, thought itself is not static and random: It develops in ways that obey different rules of inference. Haugeland says, "since correct application of the rules of reason to particular thoughts depends on what those thoughts mean, it seems that there must be some active rule-applier, which understands the thoughts (and rules), and which applies the rules to the thoughts as well as it can. If the activity of this rules applier, following the rules of reason, is to explain the rationality of our thought process, then it must be regarded as a complete little person – or homunculus (in Latin)- inside the head, directing the thoughts like a traffic cop. The trouble is: a theory that involves an homunculus to explain thinking has begged its own question, because the homunculus itself has to think, and that thinking has not been explained."

Cognitive scientists can be materialists and mentalist at the same time. They are materialist because they support the view that the mind is a complicated machines or matter. On the other hand, some support that along with mind the body exists. They can offer explanation in terms of meaning and rules following, without presupposing any unexplained homunculus. It would be peculiar to start assigning geometrical shapes and locations to the internal program routines and operation of a system. These same decisions clearly cause physical behaviour, yet no one is worried that the laws of physics are being violated. According to Haugeland, "when the machine plays, it follows rules in at least two senses: it always abides by the rules of the game, and it employs various reasonable rules of thumb to select plausible moves. Though these rules are in no way laws of nature, the machine's behaviour is explained (in part) by citing them-and yet, no unexplained 'compunculus' is presupposed." Thus this explanation will necessary invoke the system's internal reasoning processes; yet it is far from easy to figure out processes that will consistently lead to the observed behavioural response. Dennett rightly says that human mind is a semantic engine, that is to say that the way human mind handles the meaning of a word or sentence, in the same way a machine handles the literal meaning of a word or a sentence. Thus Dennett's view shows that human mind is a machine like ordinary machine because both mind and machine have the same quality, the difference is only apparent.

## Strong AI And Weak AI (Reductionism and Non-Reductionism)

As we have seen, the main thesis of AI is that the human brain is like a digital computer, and the human mind is just a computer program. It tries to prove that the relation between the programs and the computer hardware is like the relation between mind and brain. Some supporters of AI argue that we have every reason to believe that computers have intelligence. At the same time, some others argue that the computers' intelligence is limited whereas human intelligence has no limit. Nowadays computers have achieved some modest success in proving theorems, guiding missiles, sorting mail, driving assembly-line robots, diagnosing illness, predicting weather and economic events, etc. Computers receive, interpret, process, store, manipulate and use information. Thus, intelligent behaviour is programmed into the computers. On the contrary, we have no idea how the brain functions, but we have an idea of the general relationships between brain processes and mental processes. Mental processes are caused by the brain activities which are functions of the elements constituting the brain.

Strong AI argues that it is possible that one-day a computer will be invented which can function like a mind in the fullest sense of the word. In other words, it can think, reason, imagine, etc., and do all the things that we currently associate with the human minds. On the other hand, weak AI argues that computers can only simulate human mind and are not actually conscious in the same way as human minds are. According to weak AI, computers having artificial intelligence are very powerful instruments in the hands of man. Whereas Strong AI holds that computer is not merely an instrument in the study of the mind, but that the appropriately programmed computer is really a mind in the sense that computers can think and do reasoning like the human beings. In Strong AI, the programmed computer has cognitive states, so the programs are not simple tools that allow us to test psychological explanations; rather the programs are themselves the explanations. Strong AI, according to Searle, basically claims that the appropriately programmed computer literally has cognitive states, and that the programs thereby explain human cognition.

The main aim of AI is to reproduce mentality in computational machines, and to try to prove that the functions of a machine are similar to the functions of the human mind. But the question is: Could a machine have mental states? For AI, in the words of Searle, "there is nothing essentially biological about the human mind. The brain just happens to be one of an indefinitely large number of different kinds of hardware computers that could sustain the programs, which make up human intelligence. On this view, any physical system whatever that had the right program with the right inputs and outputs would have a mind in exactly the same sense that you and I have minds."

Searle is here critical of the view that any physical system that has the right program with the right inputs and outputs would have a mind in exactly the same sense that human beings have minds. The cognitive scientists believe that perhaps they can design the appropriate hardware and programs - artificial brains, and minds- that are comparable to human brains and minds. Strong artificial intelligence is a reductionist theory. Because strong AI reduces mind or mentality to physical properties. Here, the term 'reduces to' names a relation between theories. When this relation holds between a pair of theories, for example, $R^1$ and $R^2$, then $R^2$ is said to be reducer of $R^1$. According to Fodor, "the reduction relation is transitive and asymmetrical, hence irreflexible. By the 'unity of science' I shall mean the doctrine that all sciences except physics reduce to physics. By 'physicalistic reduction' I shall mean a certain claim that is entailed by, but does not entail, the unity of science; viz, the claim that psychology reduces to physics."

Reducibility involves a good deal more than the ontological claim that things that satisfy descriptions in the vocabulary of $R^1$ also satisfy descriptions the vocabulary of $R^2$. This condition is stronger than the ontological requirement that whatever falls under the generalizations of $R^1$ should also fall under those of $R^2$. On this view, there is an important sense in which syntax is preserved under reduction. That is to say, reduction permits us to redescribe the events in the vocabulary of $R^2$. Thus according to strong AI, mental states reduce to the computational states in the same way.

On the other hand, weak AI is non-reductionist, because this theory is not reducing the human mind in terms of machines, but it can only simulate human mind and this does not mean exact replication. The above statement shows that the weak AI view is non-reductionist. For the physicalist, life is a higher order property, which emerges out of the physical properties. However, in case of a zombie, there is an absence of consciousness. In other words, the logical possibility of a zombie world is considered as a world physically identical to our world, but conscious experiences is impossible in this world. The zombies may be psychological or phenomenal zombies, which are physically and functionally identical to human beings but they lack experiences. According to Chalmers, "the logical possibility of zombies seems equally obvious to me. A zombie is just something physically identical to me but which has no conscious experience – all is dark inside."

The zombie and me have identical physical properties but differ in high-level properties like consciousness. The zombies lack consciousness. Therefore, the high-level property of being conscious cannot be logically supervienent on physical properties. In the same way, according to weak AI, mind and machines have some identical properties but differ mainly on some higher qualities. Here, weak AI is non-reductionist because unlike strong AI, is does not reduce mind to machines.

## Philosophy of Artificial Intelligence

The philosophy of artificial intelligence attempts to answer such questions as follows:

- Can a machine act intelligently? Can it solve *any* problem that a person would solve by thinking?
- Are human intelligence and machine intelligence the same? Is the human brain essentially a computer?
- Can a machine have a mind, mental states, and consciousness in the same way that a human being can? Can it *feel how things are*?

These three questions reflect the divergent interests of AI researchers, Linguists, cognitive scientists and philosophers respectively. The scientific answers to these questions depend on the definition of "intelligence" and "consciousness" and exactly which "machines" are under discussion.

Important propositions in the philosophy of AI include:

- Turing's "polite convention": *If a machine behaves as intelligently as a human being, then it is as intelligent as a human being.*
- The Dartmouth proposal: *"Every aspect of learning or any other feature of intelligence can be so precisely described that a machine can be made to simulate it."*
- Newell and Simon's physical symbol system hypothesis: *"A physical symbol system has the necessary and sufficient means of general intelligent action."*
- Searle's strong AI hypothesis: *"The appropriately programmed computer with the right inputs and outputs would thereby have a mind in exactly the same sense human beings have minds."*

- Hobbes' mechanism: *"Reason is nothing but reckoning."*

## Can a Machine Display General Intelligence?

Is it possible to create a machine that can solve *all* the problems humans solve using their intelligence? This question defines the scope of what machines will be able to do in the future and guides the direction of AI research. It only concerns the *behavior* of machines and ignores the issues of interest to psychologists, cognitive scientists and philosophers; to answer this question, it does not matter whether a machine is *really* thinking (as a person thinks) or is just *acting like* it is thinking.

The basic position of most AI researchers is summed up in this statement, which appeared in the proposal for the Dartmouth workshop of 1956:

- *Every aspect of learning or any other feature of intelligence can be so precisely described that a machine can be made to simulate it.*

Arguments against the basic premise must show that building a working AI system is impossible, because there is some practical limit to the abilities of computers or that there is some special quality of the human mind that is necessary for thinking and yet cannot be duplicated by a machine (or by the methods of current AI research). Arguments in favor of the basic premise must show that such a system is possible.

The first step to answering the question is to clearly define "intelligence."

### Intelligence

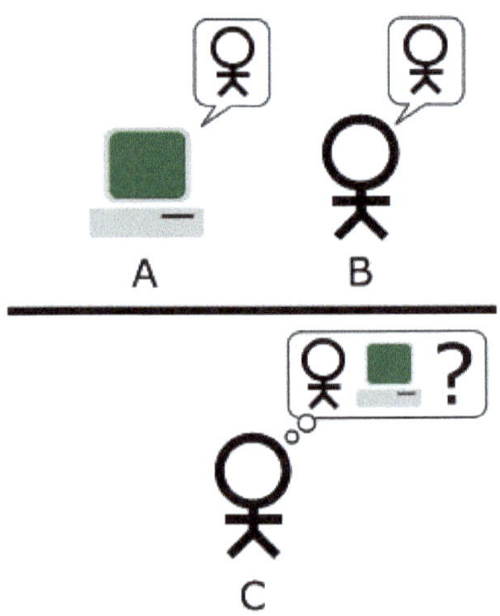


The "standard interpretation" of the Turing test.

### Turing Test

Alan Turing, in a famous and seminal 1950 paper, reduced the problem of defining intelligence to a simple question about conversation. He suggests that: if a machine can answer *any* question put to it, using the same words that an ordinary person would, then we may call that machine intelligent. A modern version of his experimental design would use an online chat room, where one of the participants is a real person and one of the participants is a computer program. The program passes the test if no one can tell which of the two participants is human. Turing notes that no one (except philosophers) ever asks the question "can people think?" He writes "instead of arguing continually over this point, it is usual to have a polite convention that everyone thinks." Turing's test extends this polite convention to machines:

- *If a machine acts as intelligently as human being, then it is as intelligent as a human being.*

One criticism of the Turing test is that it is explicitly anthropomorphic. If our ultimate goal is to create machines that are *more* intelligent than people, why should we insist that our machines must closely *resemble* people? Russell and Norvig write that "aeronautical engineering texts do not define the goal of their field as 'making machines that fly so exactly like pigeons that they can fool other pigeons.'"

## Intelligent Agent Definition

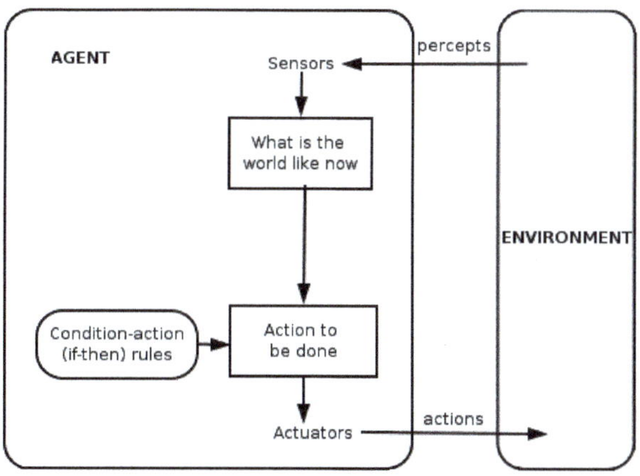


Simple reflex agent

Recent AI research defines intelligence in terms of intelligent agents. An "agent" is something which perceives and acts in an environment. A "performance measure" defines what counts as success for the agent.

- *If an agent acts so as to maximize the expected value of a performance measure based on past experience and knowledge then it is intelligent.*

Definitions like this one try to capture the essence of intelligence. They have the advantage that, unlike the Turing test, they do not also test for human traits that we may not want to consider intelligent, like the ability to be insulted or the temptation to lie. They have the disadvantage that they fail to make the commonsense differentiation between "things that think" and "things that do not". By this definition, even a thermostat has a rudimentary intelligence.

## Arguments that a Machine can Display General Intelligence

### The Brain can be Simulated

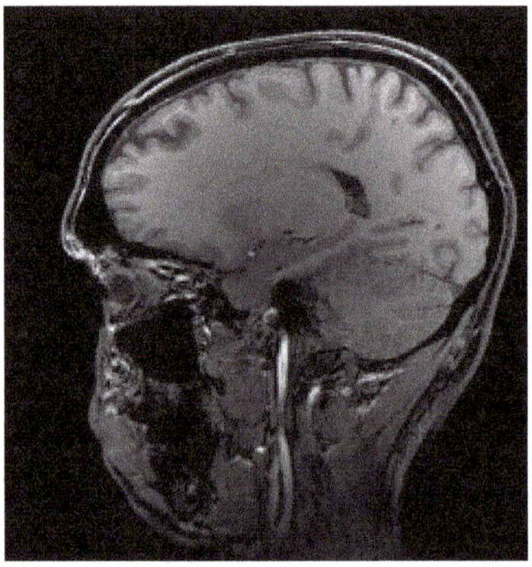

An MRI scan of a normal adult human brain

Hubert Dreyfus describes this argument as claiming that "if the nervous system obeys the laws of physics and chemistry, which we have every reason to suppose it does, then .... we ... ought to be able to reproduce the behavior of the nervous system with some physical device." This argument, first introduced as early as 1943 and vividly described by Hans Moravec in 1988, is now associated with futurist Ray Kurzweil, who estimates that computer power will be sufficient for a complete brain simulation by the year 2029. A non-real-time simulation of a thalamocortical model that has the size of the human brain ($10^{11}$ neurons) was performed in 2005 and it took 50 days to simulate 1 second of brain dynamics on a cluster of 27 processors.

Few disagree that a brain simulation is possible in theory, even critics of AI such as Hubert Dreyfus and John Searle. However, Searle points out that, in principle, *anything* can be simulated by a computer; thus, bringing the definition to its breaking point leads to the conclusion that any process at all can technically be considered "computation". "What we wanted to know is what distinguishes the mind from thermostats and livers," he writes. Thus, merely mimicking the functioning of a brain would in itself be an admission of ignorance regarding intelligence and the nature of the mind.

### Human Thinking is Symbol Processing

In 1963, Allen Newell and Herbert A. Simon proposed that "symbol manipulation" was the essence of both human and machine intelligence. They wrote:

- *A physical symbol system has the necessary and sufficient means of general intelligent action.*

This claim is very strong: it implies both that human thinking is a kind of symbol manipulation (because a symbol system is *necessary* for intelligence) and that machines can be intelligent (because a symbol system is *sufficient* for intelligence). Another version of this position was described by philosopher Hubert Dreyfus, who called it "the psychological assumption":

- *The mind can be viewed as a device operating on bits of information according to formal rules.*

A distinction is usually made between the kind of high level symbols that directly correspond with objects in the world, such as <dog> and <tail> and the more complex "symbols" that are present in a machine like a neural network. Early research into AI, called "good old fashioned artificial intelligence" (GOFAI) by John Haugeland, focused on these kind of high level symbols.

## Arguments Against Symbol Processing

These arguments show that human thinking does not consist (solely) of high level symbol manipulation. They do *not* show that artificial intelligence is impossible, only that more than symbol processing is required.

## Gödelian Anti-mechanist Arguments

In 1931, Kurt Gödel proved with an incompleteness theorem that it is always possible to construct a "Gödel statement" that a given consistent formal system of logic (such as a high-level symbol manipulation program) could not prove. Despite being a true statement, the constructed Gödel statement is unprovable in the given system. (The truth of the constructed Gödel statement is contingent on the consistency of the given system; applying the same process to a subtly inconsistent system will appear to succeed, but will actually yield a false "Gödel statement" instead.) More speculatively, Gödel conjectured that the human mind can correctly eventually determine the truth or falsity of any well-grounded mathematical statement (including any possible Gödel statement), and that therefore the human mind's power is not reducible to a *mechanism*. Philosopher John Lucas (since 1961) and Roger Penrose (since 1989) have championed this philosophical anti-mechanist argument. Gödelian anti-mechanist arguments tend to rely on the innocuous-seeming claim that a system of human mathematicians (or some idealization of human mathematicians) is both consistent (completely free of error) and believes fully in its own consistency (and can make all logical inferences that follow from its own consistency, including belief in its Gödel statement). This is provably impossible for a Turing machine (and, by an informal extension, any known type of mechanical computer) to do; therefore, the Gödelian concludes that human reasoning is too powerful to be captured in a machine.

However, the modern consensus in the scientific and mathematical community is that actual human reasoning is inconsistent; that any consistent "idealized version" *H* of human reasoning would logically be forced to adopt a healthy but counter-intuitive open-minded skepticism about the consistency of *H* (otherwise *H* is provably inconsistent); and that Gödel's theorems do not lead to any valid argument that humans have mathematical reasoning capabilities beyond what a machine could ever duplicate. This consensus that Gödelian anti-mechanist arguments are doomed to failure is laid out strongly in *Artificial Intelligence*: "*any* attempt to utilize (Gödel's incompleteness results) to attack the computationalist thesis is bound to be illegitimate, since these results are quite consistent with the computationalist thesis."

More pragmatically, Russell and Norvig note that Gödel's argument only applies to what can theoretically be proved, given an infinite amount of memory and time. In practice, real machines (including humans) have finite resources and will have difficulty proving many theorems. It is not necessary to prove everything in order to be intelligent.

Less formally, Douglas Hofstadter, in his Pulitzer prize winning book *Gödel, Escher, Bach: An Eternal Golden Braid,* states that these "Gödel-statements" always refer to the system itself, drawing an analogy to the way the Epimenides paradox uses statements that refer to themselves, such as "this statement is false" or "I am lying". But, of course, the Epimenides paradox applies to anything that makes statements, whether they are machines *or* humans, even Lucas himself. Consider:

- *Lucas can't assert the truth of this statement.*

This statement is true but cannot be asserted by Lucas. This shows that Lucas himself is subject to the same limits that he describes for machines, as are all people, and so Lucas's argument is pointless.

After concluding that human reasoning is non-computable, Penrose went on to controversially speculate that some kind of hypothetical non-computable processes involving the collapse of quantum mechanical states give humans a special advantage over existing computers. Existing quantum computers are only capable of reducing the complexity of Turing computable tasks and are still restricted to tasks within the scope of Turing machines. Quantum computer - relation to computational complexity theory. By Penrose and Lucas's arguments, existing quantum computers are not sufficient, so Penrose seeks for some other process involving new physics, for instance quantum gravity which might manifest new physics at the scale of the Plank mass via spontaneous quantum collapse of the wave function. These states, he suggested, occur both within neurons and also spanning more than one neuron. However, other scientists point out that there is no plausible organic mechanism in the brain for harnessing any sort of quantum computation, and furthermore that the timescale of quantum decoherence seems too fast to influence neuron firing.

## Dreyfus: The Primacy of Unconscious Skills

Hubert Dreyfus argued that human intelligence and expertise depended primarily on unconscious instincts rather than conscious symbolic manipulation, and argued that these unconscious skills would never be captured in formal rules.

Dreyfus's argument had been anticipated by Turing in his 1950 paper Computing machinery and intelligence, where he had classified this as the "argument from the informality of behavior." Turing argued in response that, just because we do not know the rules that govern a complex behavior, this does not mean that no such rules exist. He wrote: "we cannot so easily convince ourselves of the absence of complete laws of behaviour ... The only way we know of for finding such laws is scientific observation, and we certainly know of no circumstances under which we could say, 'We have searched enough. There are no such laws.'"

Russell and Norvig point out that, in the years since Dreyfus published his critique, progress has been made towards discovering the "rules" that govern unconscious reasoning. The situated movement in robotics research attempts to capture our unconscious skills at perception and attention. Computational intelligence paradigms, such as neural nets, evolutionary algorithms and so on are mostly directed at simulated unconscious reasoning and learning. Statistical approaches to AI can make predictions which approach the accuracy of human intuitive guesses. Research into commonsense knowledge has focused on reproducing the "background" or context of knowledge. In fact, AI research in general has moved away from high level symbol manipulation or "GOFAI", towards new models that are intended to capture more of our *unconscious* reasoning. Historian and AI researcher Daniel Crevier wrote that "time has proven the accuracy and perceptiveness of some of Dreyfus's comments. Had he formulated them less aggressively, constructive actions they suggested might have been taken much earlier."

## Can a Machine have a Mind, Consciousness, and Mental States?

This is a philosophical question, related to the problem of other minds and the hard problem of consciousness. The question revolves around a position defined by John Searle as "strong AI":

- *A physical symbol system can have a mind and mental states.*

Searle distinguished this position from what he called "weak AI":

- *A physical symbol system can act intelligently.*

Searle introduced the terms to isolate strong AI from weak AI so he could focus on what he thought was the more interesting and debatable issue. He argued that *even if we assume* that we had a computer program that acted exactly like a human mind, there would still be a difficult philosophical question that needed to be answered.

Neither of Searle's two positions are of great concern to AI research, since they do not directly answer the question "can a machine display general intelligence?" (unless it can also be shown that consciousness is *necessary* for intelligence). Turing wrote "I do not wish to give the impression that I think there is no mystery about consciousness... [b]ut I do not think these mysteries necessarily need to be solved before we can answer the question [of whether machines can think]." Russell and Norvig agree: "Most AI researchers take the weak AI hypothesis for granted, and don't care about the strong AI hypothesis."

There are a few researchers who believe that consciousness is an essential element in intelligence, such as Igor Aleksander, Stan Franklin, Ron Sun, and Pentti Haikonen, although their definition of "consciousness" strays very close to "intelligence."

Before we can answer this question, we must be clear what we mean by "minds", "mental states" and "consciousness".

## Consciousness, Minds, Mental States, Meaning

The words "mind" and "consciousness" are used by different communities in different ways. Some new age thinkers, for example, use the word "consciousness" to describe something similar to Bergson's "élan vital": an invisible, energetic fluid that permeates life and especially the mind. Science fiction writers use the word to describe some essential property that makes us human: a machine or alien that is "conscious" will be presented as a fully human character, with intelligence, desires, will, insight, pride and so on. (Science fiction writers also use the words "sentience", "sapience," "self-awareness" or "ghost" - as in the *Ghost in the Shell* manga and anime series - to describe this essential human property). For others, the words "mind" or "consciousness" are used as a kind of secular synonym for the soul.

For philosophers, neuroscientists and cognitive scientists, the words are used in a way that is both more precise and more mundane: they refer to the familiar, everyday experience of having a "thought in your head", like a perception, a dream, an intention or a plan, and to the way we *know* something, or *mean* something or *understand* something. "It's not hard to give a commonsense definition of consciousness" observes philosopher John Searle. What is mysterious and fascinating is not so much *what* it is but *how* it is: how does a lump of fatty tissue and electricity give rise to this (familiar) experience of perceiving, meaning or thinking?

Philosophers call this the hard problem of consciousness. It is the latest version of a classic problem in the philosophy of mind called the "mind-body problem." A related problem is the problem of *meaning* or *understanding* (which philosophers call "intentionality"): what is the connection between our *thoughts* and *what we are thinking about* (i.e. objects and situations out in the world)? A third issue is the problem of *experience* (or "phenomenology"): If two people see the same thing, do they have the same experience? Or are there things "inside their head" (called "qualia") that can be different from person to person?

Neurobiologists believe all these problems will be solved as we begin to identify the neural correlates of consciousness: the actual relationship between the machinery in our heads and its collective properties; such as the mind, experience and understanding. Some of the harshest critics of artificial intelligence agree that the brain is just a machine, and that consciousness and intelligence are the result of physical processes in the brain. The difficult philosophical question is this: can a computer program, running on a digital machine that shuffles the binary digits of zero and one, duplicate the ability of the neurons to create minds, with mental states (like understanding or perceiving), and ultimately, the experience of consciousness?

## Arguments that a Computer Cannot have a Mind and Mental States

### Searle's Chinese Room

John Searle asks us to consider a thought experiment: suppose we have written a computer program that passes the Turing test and demonstrates "general intelligent action." Suppose, specifically that the program can converse in fluent Chinese. Write the program on 3x5 cards and give them to an ordinary person who does not speak Chinese. Lock the person into a room and have him follow the instructions on the cards. He will copy out Chinese characters and pass them in and out of the room through a slot. From the outside, it will appear that the Chinese room contains a fully intelligent person who speaks Chinese. The question is this: is there anyone (or anything) in the room that understands Chinese? That is, is there anything that has the mental state of understanding, or which has conscious awareness of what is being discussed in Chinese? The man is clearly not aware. The room cannot be aware. The *cards* certainly aren't aware. Searle concludes that the Chinese room, or *any* other physical symbol system, cannot have a mind.

Searle goes on to argue that actual mental states and consciousness require (yet to be described) "actual physical-chemical properties of actual human brains." He argues there are special "causal properties" of brains and neurons that gives rise to minds: in his words "brains cause minds."

### Related Arguments: Leibniz' Mill, Davis's Telephone Exchange, Block's Chinese Nation and Blockhead

Gottfried Leibniz made essentially the same argument as Searle in 1714, using the thought experiment of expanding the brain until it was the size of a mill. In 1974, Lawrence Davis imagined duplicating the brain using telephone lines and offices staffed by people, and in 1978 Ned Block envisioned the entire population of China involved in such a brain simulation. This thought experiment is called "the Chinese Nation" or "the Chinese Gym". Ned Block also proposed his "blockhead" argument, which is a version of the Chinese room in which the program has been re-factored into a simple set of rules of the form "see this, do that", removing all mystery from the program.

## Responses to the Chinese Room

Responses to the Chinese room emphasize several different points.

- The systems reply and the virtual mind reply: This reply argues that the system, including the man, the program, the room, and the cards, is what understands Chinese. Searle claims that the man in the room is the only thing which could possibly "have a mind" or "understand", but others disagree, arguing that it is possible for there to be two minds in the same physical place, similar to the way a computer can simultaneously "be" two machines at once: one physical (like a Macintosh) and one "virtual" (like a word processor).
- Speed, power and complexity replies: Several critics point out that the man in the room would probably take millions of years to respond to a simple question, and would require "filing cabinets" of astronomical proportions. This brings the clarity of Searle's intuition into doubt.
- Robot reply: To truly understand, some believe the Chinese Room needs eyes and hands. Hans Moravec writes: 'If we could graft a robot to a reasoning program, we wouldn't need a person to provide the meaning anymore: it would come from the physical world."
- Brain simulator reply: What if the program simulates the sequence of nerve firings at the synapses of an actual brain of an actual Chinese speaker? The man in the room would be simulating an actual brain. This is a variation on the "systems reply" that appears more plausible because "the system" now clearly operates like a human brain, which strengthens the intuition that there is something besides the man in the room that could understand Chinese.
- Other minds reply and the epiphenomena reply: Several people have noted that Searle's argument is just a version of the problem of other minds, applied to machines. Since it is difficult to decide if people are "actually" thinking, we should not be surprised that it is difficult to answer the same question about machines.

  A related question is whether "consciousness" (as Searle understands it) exists. Searle argues that the experience of consciousness can't be detected by examining the behavior of a machine, a human being or any other animal. Daniel Dennett points out that natural selection cannot preserve a feature of an animal that has no effect on the behavior of the animal, and thus consciousness (as Searle understands it) can't be produced by natural selection. Therefore either natural selection did not produce consciousness, or "strong AI" is correct in that consciousness can be detected by suitably designed Turing test.

## Is Thinking a Kind of Computation?

The computational theory of mind or "computationalism" claims that the relationship between mind and brain is similar (if not identical) to the relationship between a *running program* and a computer. The idea has philosophical roots in Hobbes (who claimed reasoning was "nothing more than reckoning"), Leibniz (who attempted to create a logical calculus of all human ideas), Hume (who thought perception could be reduced to "atomic impressions") and even Kant (who analyzed all experience as controlled by formal rules). The latest version is associated with philosophers Hilary Putnam and Jerry Fodor.

This question bears on our earlier questions: if the human brain is a kind of computer then computers can be both intelligent and conscious, answering both the practical and philosophical questions of AI. In terms of the practical question of AI ("Can a machine display general intelligence?"), some versions of computationalism make the claim that (as Hobbes wrote):

- *Reasoning is nothing but reckoning*

In other words, our intelligence derives from a form of *calculation*, similar to arithmetic. This is the physical symbol system hypothesis discussed above, and it implies that artificial intelligence is possible. In terms of the philosophical question of AI ("Can a machine have mind, mental states and consciousness?"), most versions of computationalism claim that (as Stevan Harnad characterizes it):

- *Mental states are just implementations of (the right) computer programs*

This is John Searle's "strong AI" discussed above, and it is the real target of the Chinese room argument (according to Harnad).

## Other Related Questions

Alan Turing noted that there are many arguments of the form "a machine will never do X", where X can be many things, such as:

Be kind, resourceful, beautiful, friendly, have initiative, have a sense of humor, tell right from wrong, make mistakes, fall in love, enjoy strawberries and cream, make someone fall in love with it, learn from experience, use words properly, be the subject of its own thought, have as much diversity of behaviour as a man, do something really new.

Turing argues that these objections are often based on naive assumptions about the versatility of machines or are "disguised forms of the argument from consciousness". Writing a program that exhibits one of these behaviors "will not make much of an impression." All of these arguments are tangential to the basic premise of AI, unless it can be shown that one of these traits is essential for general intelligence.

## Can a Machine have Emotions?

If "emotions" are defined only in terms of their effect on behavior or on how they function inside an organism, then emotions can be viewed as a mechanism that an intelligent agent uses to maximize the utility of its actions. Given this definition of emotion, Hans Moravec believes that "robots in general will be quite emotional about being nice people". Fear is a source of urgency. Empathy is a necessary component of good human computer interaction. He says robots "will try to please you in an apparently selfless manner because it will get a thrill out of this positive reinforcement. You can interpret this as a kind of love." Daniel Crevier writes "Moravec's point is that emotions are just devices for channeling behavior in a direction beneficial to the survival of one's species."

However, emotions can also be defined in terms of their subjective quality, of what it *feels like* to have an emotion. The question of whether the machine *actually feels* an emotion, or whether it merely *acts as if* it is feeling an emotion is the philosophical question, "can a machine be conscious?" in another form.

## Can a Machine be Self-aware?

"Self awareness", as noted above, is sometimes used by science fiction writers as a name for the essential human property that makes a character fully human. Turing strips away all other properties of human beings and reduces the question to "can a machine be the subject of its own thought?" Can it *think about itself*? Viewed in this way, it is obvious that a program can be written that can report on its own internal states, such as a debugger. Though arguably self-awareness often presumes a bit more capability; a machine that can ascribe meaning in some way to not only its own state but in general postulating questions without solid answers: the contextual nature of its existence now; how it compares to past states or plans for the future, the limits and value of its work product, how it perceives its performance to be valued-by or compared to others.

## Can a Machine be Original or Creative?

Turing reduces this to the question of whether a machine can "take us by surprise" and argues that this is obviously true, as any programmer can attest. He notes that, with enough storage capacity, a computer can behave in an astronomical number of different ways. It must be possible, even trivial, for a computer that can represent ideas to combine them in new ways. (Douglas Lenat's Automated Mathematician, as one example, combined ideas to discover new mathematical truths.)

In 2009, scientists at Aberystwyth University in Wales and the U.K's University of Cambridge designed a robot called Adam that they believe to be the first machine to independently come up with new scientific findings. Also in 2009, researchers at Cornell developed Eureqa, a computer program that extrapolates formulas to fit the data inputted, such as finding the laws of motion from a pendulum's motion.

## Can a Machine be Benevolent or Hostile?

This question (like many others in the philosophy of artificial intelligence) can be presented in two forms. "Hostility" can be defined in terms function or behavior, in which case "hostile" becomes synonymous with "dangerous". Or it can be defined in terms of intent: can a machine "deliberately" set out to do harm? The latter is the question "can a machine have conscious states?" (such as intentions) in another form.

The question of whether highly intelligent and completely autonomous machines would be dangerous has been examined in detail by futurists (such as the Singularity Institute). (The obvious element of drama has also made the subject popular in science fiction, which has considered many differently possible scenarios where intelligent machines pose a threat to mankind.)

One issue is that machines may acquire the autonomy and intelligence required to be dangerous very quickly. Vernor Vinge has suggested that over just a few years, computers will suddenly become thousands or millions of times more intelligent than humans. He calls this "the Singularity." He suggests that it may be somewhat or possibly very dangerous for humans. This is discussed by a philosophy called Singularitarianism.

In 2009, academics and technical experts attended a conference to discuss the potential impact of robots and computers and the impact of the hypothetical possibility that they could become self-sufficient and able to make their own decisions. They discussed the possibility and the extent to which computers and robots might be able to acquire any level of autonomy, and to what degree they could use such abilities to possibly pose any threat or hazard. They noted that some machines have acquired various forms of semi-autonomy, including being able to find power sources on their own and being able to independently choose targets to attack with weapons. They also noted that some computer viruses can evade elimination and have achieved "cockroach intelligence." They noted that self-awareness as depicted in science-fiction is probably unlikely, but that there were other potential hazards and pitfalls.

Some experts and academics have questioned the use of robots for military combat, especially when such robots are given some degree of autonomous functions. The US Navy has funded a report which indicates that as military robots become more complex, there should be greater attention to implications of their ability to make autonomous decisions.

The President of the Association for the Advancement of Artificial Intelligence has commissioned a study to look at this issue. They point to programs like the Language Acquisition Device which can emulate human interaction.

Some have suggested a need to build "Friendly AI", meaning that the advances which are already occurring with AI should also include an effort to make AI intrinsically friendly and humane.

## Can a Machine have a Soul?

Finally, those who believe in the existence of a soul may argue that "Thinking is a function of man's immortal soul." Alan Turing called this "the theological objection" and considers it on its own merits. He writes

In attempting to construct such machines we should not be irreverently usurping His power of creating souls, any more than we are in the procreation of children: rather we are, in either case, instruments of His will providing mansions for the souls that He creates.

## Conclusion and Themes for Future Research

John McCarthy, who created the LISP programming language for AI, and AI concept itself, says, that some philosophers of AI will do battle with the idea that:

- AI is impossible (Dreyfus),
- AI is immoral (Weizenbaum),
- The very concept of AI is incoherent (Searle).

## The General Argument Against AI.

Over the past decades, electronics and computer technology has made great strides in the sphere of knowledge and has helped us in our dealing with the world. The computers of today are much more developed and sophisticated than the mechanical calculators of yesterday. Already computers are able to perform numerous tasks that had previously been the exclusive province of human beings, with a speed and accuracy that far outstrip anything that a human being can achieve.

Moreover, the advent of computer technology has given a new direction to our understanding of intelligence, thought and other mental activities. We are inclined to raise such questions like: What does it mean to think or to feel? What is mind? Does mind really exist? Besides, we may raise the questions: To what extent are minds functionally dependent upon the physical structures with which they are associated? Are minds subject to the law of physics? If so, what are the laws of physics? Of course, to ask for definite answers to such questions would be a tall order. These questions are eminently philosophical in nature. In philosophy of mind we are interested in understanding the nature of mind, thought, intelligence, etc. as it enables us to appreciate the notions of machine-mind and machine-intelligence.

However, the idea of machine-intelligence has been challenged by philosophers and logicians in recent times. There are a number of results of mathematical logic, which can be used to show that there are limitations to the powers of discrete state machines. The best known of these results is known as Gödel's theorem, which shows that in any sufficiently powerful logical system statements can be formulated which can neither be proved nor disproved within the system, unless possibly the system itself is inconsistent. As Lucas says, “Gödel’s theorem must apply to cybernetical machines, because it is of the essence of being a machine, that it should be a concrete instantiation of a formal system. It follows that given any machine which is consistent and capable of doing simple arithmetic, there is a formula which is incapable of producing as being true- i.e., the formula is unprovable-in-the-system -but which we can see to be true. It follows that no machine can be a complete or adequate model of the mind, that minds are essentially different from machines.”

According to Gödel, no logical system can be held to be self-complete, so it always needs another system to prove its completeness. Now, in the light of this, it can be shown that no machine, like a logical system, can be self-complete, that is, cannot do everything. That is to say, there will be some questions to which it will either give a wrong answer, or fail to give an answer at all, however, much time is allowed for a reply. There may be many such questions, that cannot be answered by one machine, but may be satisfactorily answered by another. Thus machines have limitations in their functions. Therefore, Turing says, “I grant you that you can make machines do all the things you have mentioned but you will never be able to make them do X.” He mentioned numerous features of X in this connection, which could not be performed by a machine. These features are mentioned by him in the following passage:

“Be kind, resourceful, beautiful, friendly, have initiative, have a sense of humor, tell right from wrong, make mistakes, enjoy strawberries and cream, make someone fall in love with it, learn from experience, use words properly, be the subject of its own thought, have as much diversity of behaviour as a man do something really new.” These are the various disability arguments

A man uses a number of machines in his lifetime. The general belief that machines cannot commit mistakes is not always true. Machines do go wrong. Turing’s notion of imitation game may be taken into account. In this game machines would be given a set of problems to solve. One would deliberately introduce mistakes in a manner calculated to confuse the machine. As a result, a mechanical fault would probably show itself through introduction of a wrong move in the calculation. Thus machines are vulnerable to all kinds of mistakes.

The above criticism of machines shows two kinds of mistakes that the machines can commit. We may call them ‘errors of functioning’ and ‘errors of conclusion.’ Errors of functioning are due to some mechanical or electrical faults, which cause the machine to behave otherwise than it is designed to do. In philosophical discussions, one likes to ignore the possibility of such errors, because we are discussing ‘abstract machines.’ These abstract machines are mathematical fictions rather than physical objects. By definition they are incapable of errors of functioning. In this sense, we can say that, machines can never make mistakes. However, the machines can commit errors of conclusion because they can make mistake moves in this function. These mistaken are the errors of argument.

According to Lovelace, "The Analytical Engine has no pretensions to originate anything. It can do whatever we know how to order it to perform." Here, the analytical Engine referred to is a universal digital computer, which can be made to perform many task but cannot originate anything on its own. That is to say, as a machine it fails in the matter of creativity. It is for this reason we can say that a machine can 'never do anything really new.' In comparison to the machine, we can argue, the human mind is not a machine at all, since it originates many new things. The human system as a whole is a creative system which cannot be mimicked by any machine.

Again, we may argue that human beings have some psychological qualities such as intelligence, consciousness or originality, etc. which are said to be necessarily lacking in machine. That is why, a machine is normally treated as an artefact and a mere mechanical contrivance manufactured for a definite purpose.

However, when it is said that it is impossible for a machine to be conscious, it is not always clear to what extent this is intended to be a logical objection, and to what extent empirical. Empirically, machines are not conscious, but this cannot be proved logically. Robots are well known for duplicating human behaviour. For a robot, ex-hypothesis, is capable of behaving like a human being. We have no doubt that a human being is conscious, when he or she is doing work. Though a machine might do the same work, we are not inclined call the latter conscious. Thus it is taken for granted that humans are conscious, whereas of machines we enquire whether they are capable of consciousness or not. We know that the question of consciousness is appropriate in the context of human beings, but not so in the case of machines. A machine is essentially distinct from a man so far as consciousness is concerned. The machine-intelligence and machine-behaviour are not indicative of consciousness at all.

Michal Scriven has raised the question as to whether a robot can be considered as conscious. He is of the view that a robot is simply a machine which is indistinguishable from humans in behavioural aspects. In spite of the close similarity in behaviour, however, human and robots are essentially different, belonging to two different types of entities. Campbell, therefore, has argued, following Descartes, that human beings alone and not machines are conscious. Machines are unconscious material bodies while human beings are conscious entities.

Here, a question may arise: Is it a blind prejudice which prevents us from extending the attribute of consciousness to robots, when robots can calculate more quickly, react more swiftly, see more clearly, and remember more accurately than men? What is it that they lack when they can do so many things? They do what humans do, yet they cannot be treated at par with human beings. The obvious answer to this is that the robots have no consciousness. They are only machines imitating human beings.

As Michael Scriven points out, the sense of 'conscious' is contrasted with 'incapable of being conscious' and we can ascribe consciousness to creatures only if we can also withhold such ascription to them. He says that it is absurd to ask of a stone or a stop-watch 'Is it conscious?' because it is absurd to talk of its being dead, asleep, drugged or stunned, i.e., unconscious. There are cases where it is very difficult to decide the question of consciousness. For example, let us take a man who has a completely anaesthetized cortex, and who has an external operator controlling his or her outer behaviour. In such cases, the man is unconscious, though outwardly conscious. But such difficulties arise only about conscious beings. In this case, the man was conscious but now has no consciousness, his brain beings out of action temporarily.

It is now certain that under no circumstances can we prove that the robot is conscious. We have a complete causal explanation of all its behaviour, and this explanation does not at any stage depend on its consciousness. So its behaviour cannot be a proof of the possession of consciousness. Following the above argument, Scriven holds that consciousness is not a property, which can be detected in a machine by any physical examination, because it cannot be identified with any physical characteristic of a machine.

Therefore, even if a computer does exactly what a human being does, it can never be ascribed consciousness. It never does anything creative or new or unpredictable. Its output is the result of its physical structure, its program, and the input it is given. A human being, on the other hand, initiates novel, creative and unpredictable actions. Thus a human being stands on a different footing from the computer. This argument can be laid against AI, since there is a wide logical gap between human beings and the computing machines. Computers not only lack creativity but they lack basic capacity to learn. Many people take unpredictability as an evidence for originality, and fear that if it is true that mentality bottoms out in straight forwardly mechanical processes, we eventually will be able to predict everything about people. And at that point human life will lose its joy and mystery. Hence we can argue that people, and not machines have creativity.

In defence of AI, however, one may say that the computer is an ideal model of mentality, because it is an ideal model of the brain. But the computer, in practice, does not simulate all the functions of the brain, and so remains an incomplete model. The Von Neumann devices also have little in common with brains. Even the claim that connectionist machines are biologically realistic requires a good deal of charity. They are more like brains than the Von Neomann devices are, but they are a lot less like brains than other brains are. Thus it can be said that computers will never come to possess genuine mentality because they do not have the human brain.

Proponents of weak psychological AI claim that we can write programs that test the relative plausibility of different psychological hypotheses. For example, one can write programs that purport to describe the cognitive mechanisms underlying language production. The correct programs will be the ones which pass the test of descriptive adequacy. However, the proponents of AI would not be able to claim that computers can accurately simulate human behaviour because they do not posses the competence of the human brains to produce conscious activity. Despite the differences between computers and brains, there is no reason to think that computers cannot represent any relevant information we desire about neural processes. The point is that a computational system can simulate a brain system without being just like the brain.

According to J.R. Lucas, Gödel's theorem states that in any consistent system, which is strong enough to produce simple arithmetic, there are formulas which cannot be proved-in-the-system, but which we can see to be true. Such a formula which is the, "This formula is unprovable-in-the-system." If this formula were provable-in-the-system, then it will be unprovable-in-the-system. So there will be a contradiction. So the formula 'This formula is unprovable-in-the-system' is not provable-in-the-system, but unprovable-in-the-system. Further, if the formula is 'This formula is unprovable-in-the-system', then it is true that the formula is unprovable-in-the-system, that is, 'This formula is unprovable-in-the-system' is true.

The whole effort of Gödel's theorem is to show that all formal systems which are (i) consistent, (ii) adequate for simple arithmetic, i.e., contain the natural numbers and the operations of addition and multiplication, and that (iii) they are incomplete, i.e., contain unprovable, though perfectly meaningful, formulae, which we can see to be true, standing outside the system.

According to J.R.Lucas, Gödel's theorem must apply to cybernetical machines, because it is of the essence of being a machine that it should be a concrete instantiation of a formal system. It follows that given any machine that is consistent and capable of doing simple arithmetic, there is a formula, which, though true, is not provable in the formal system of the machine. Thus it follows that no machine can be a complete or adequate model of the mind; that is, the minds are essentially different from machines.

As we know, a cybernetical machine is a device, which performs a set of operations according to a definite set of rules. Normally we 'program' a machine, that is, we give it a set of instructions about its functioning and we feed in the initial 'information' on which the machine is to perform its calculations. When we consider the mind on the model of cybernetical mechanism, we have a similar model in view. If human mind is such a model, mind is determined by the way it is made. Then, there is no possibility of its acting on its own, as it is governed by certain rules of construction and certain input of information. But this is not the characteristic of mind, as the mind does not act under ready-made rules.

In the machine, there are some formal rules of inference having been applied to some previous formula. According J.R. Lucas, we can construct a Gödelian formula in this formal system. This formula cannot be proved-in-the-system. Thus the machine cannot prove the corresponding formula as true. But one can see that the Gödelian formula is true. We can now see that any mechanical model of the mind must include a mechanism that can elucidate truths of arithmetic, because this is something, which minds can do. In fact, it is easy to produce mechanical models which will in many respects produce truths of arithmetic far better that what the human beings can do, but for every machine there is a truth which it can not prove, but which can be proved by the mind. Thus in the words of Lucas, "This is not to say that we cannot build a machine to simulate any desired piece of mind-like behaviour: it is only that we can not build a machine to simulate every piece of mind-like behaviour. We can (or shall be able to one day) build machines capable of reproducing bits of mind-like behaviour, and indeed of outdoing the performances of human minds: but however good the machine is, and however much better it can do in nearly all respects than a human mind can, ii always has this one weakness, this one thing which it cannot do, where as a mind can."

Moreover, Gödels argument shows that the mechanical model of mind, because of its inherent limitations cannot simulate the functions of the mind which are infinite and indefinite. Further, it shows that machines are finitely closed and hence cannot be compared with human minds.

## Searle's Argument Against AI.

Searle's main intention is to criticize and overcome the dominant traditions in the study of minds, both 'materialist' and 'dualist'. For him, consciousness is central to the mental phenomena. We think of ourselves as conscious, mindful, rational agents in the world, but science tells us that the world consists entirely of mindless physical particles. But, the question is: How can we match these two conceptions? According to Searle, can it be the case that the world contains nothing but unconscious physical particles, and yet that it also contains consciousness? Can an essentially meaningless world contain meanings? Searle writes, "I believe that the mind-body problem has a rather simple solution, one that is consistent both with what we know always neurophysiology and with our commonsense conception of the nature of mental states –pains, beliefs, desires and so on. But before presenting that solution, I want to ask why the mind-body problem seems so intractable. Why do we still have in philosophy and psychology, after all these centuries, a 'mind-body problem' in a way that we do not have, say, a 'digestion-stomach problem'? Why does the mind seem more mysterious than other biological phenomena?"

Moreover, for Searle, all the above problems spill over into other contemporary materialistic interpretations of the issues of mind. Materialism asks the question: How should we interpret the recent work in computer science and artificial intelligence aimed at making intelligent machines? More particularly, does the digital computer give us the right picture of the human mind? Thus the central issue is: What is the relation between the ordinary, commonsense explanations of people's behaviour and its scientific modes of explanation? Searle seeks to answer this question in his attack on materialism in his philosophy of mind.

Searle offers a biological explanation of mind according to which mind is a biological offshoot of the brain. In order to distinguish this view from others in the field, Searle calls it "biological naturalism." Mental events and processes are as much part of our biological natural history as digestion, mitosis, meiosis, or enzyme secretion, says Searle.

The biological naturalism raises many questions of its own. But one of the fundamental questions is: What about the great variety of our mental life-pains, desires, tickles, thoughts, visual experiences, beliefs, tastes, smell, anxiety, fear, love, hate, depression and elation? Again, some of the philosophical questions, which were raised by Searle, are: What exactly is consciousness and how exactly do conscious mental phenomena relate to the unconscious? What are the special features of the 'mental', phenomena such as consciousness, intentionality, subjectivity, and mental causation? And how exactly do they function? What are the causal relations between 'mental' phenomena and 'physical' phenomena? And can we characterize those causal relations in a way that avoids epiphenomenalism?

Searle's biological naturalism provides an effective counter argument to the currently fashionable computational theory of mind according to which, the mind is a computer program. According to this theory, the mind is to the brain what the program is to the hardware. In short, minds are computer programs implemented in brains. In Searle's words: "The brain is just a digital computer and the mind is just a computer program. One could summarize this view- I call it 'strong artificial intelligence' or 'strong-AI'-by saying that the mind is to the brain, as the program is to the computer hardware."

The notion of strong AI is called by Dennett 'computer functionalism'. Both the discipline of artificial intelligence and the philosophical theory of functionalism converge on the idea that the mind is just a computer program. For both the theories the human mind is a computational system that realizes programs, that is, it is a formal device. That produces functions of various kinds called the mental functions. It is a system which functions with the right inputs and outputs so that the resulting activities are treated as the mental activities. The supporters of strong AI say that there is a general agreement among them that it is only a matter of time until the computer scientists and the workers in artificial intelligence design the appropriate hardware and programs, which will be the equivalent of human brains and minds. These will be artificial brains and minds, which are in every way the equivalents of human brains and minds. As Herbert Simon says, "We already have machines that can literally think. There is no question of waiting for some future machine, because existing digital computers already have the same sense that you and I do." That is, the idea of a thinking machine is no more a dream but a reality. Hence, the legitimacy of strong artificial intelligence.

Alan Newell holds a similar view when he says, "we have now discovered that intelligence is just a matter of physical symbol manipulation; it has no essential connection with any specific kind of biological or physical wetware or hardware. Rather, any system whatever that is capable of manipulating physical symbols in the right way is capable of intelligence in the same literal sense as human intelligence of human beings." Marvin Minisky holds that, "the next generation of computers will be so intelligent that we will be lucky if they are willing to keep us around the house as household pets." McCarthy says that even "machines as simple as thermostats can be said to have beliefs." All these declarations by eminent scientists justify the idea of strong AI.

Searle, however, refutes the very idea of strong AI. His argument against AI has nothing to do with any particular stage of computer technology. It is important to emphasize this point because the temptation is always to think that the solution to our problems must wait on some as yet uncreated technological wonder. This refutation has to do with a definition of digital computers, and the idea of artificial intelligence underlying it.

As we know, the conception of a digital computer is that its operations can be specified purely formally; that is, we can specify the steps in the operation of the computer in terms of abstract symbols-sequences of zeros and ones printed on a tape. But the symbols have no meaning; they have no semantic content, they are not about anything. They have to be specified purely in terms of their formal or syntactical structure. By definition, our internal mental states have certain sorts of contents. Searle says, "In a word, the mind has more than a syntax, it has a semantics. The reason that no computer program can ever be a mind is simply that a computer program is only syntactical, and minds are more than syntactical. Minds are semantic in the sense that they have more than a formal structure, they have content." Searle presents a thought experiment about a Chinese Room for refuting the possibility of AI. This is called the Chinese Room Argument. He ask us to imagine that the computer programmers have written a program that will enable a computer to simulate the understanding of Chinese. Thus, for example, if the computer is given a question in Chinese, it will match the question with its memory, or database, and produce appropriate answers to the questions in Chinese. Suppose that the computer's answers are as good as those of a native Chinese speaker.

Then the question is: does the computer literally understand Chinese in the way the Chinese speakers understand Chinese? Again, let us imagine that someone is locked in a room, with several baskets full of Chinese symbols. However, let's imagine that he or she does not understand a word of Chinese and he or she is given a rulebook in English for manipulating these Chinese symbols. The rules specify the manipulations of the symbols purely formally, that is, in terms of their syntax, but not their semantics. So the rules might say, 'take a squiggle-squiggle sign out of basket number one and put it next to a squiggle squiggle sign from basket number two.' Now suppose that some other Chinese symbols are passed into room, and he is given further rules for passing back Chinese symbols out of the room. Supposing that unknown to him, the symbols passed into the room are called the 'questions' by the people outside the room, and the symbols he passes back out of the room are called 'answers to the question'. Furthermore, the programmers are so good at designing the programs that the person in the Chinese Room can easily manipulate symbol so that very soon the answers are indistinguishable from those of a native Chinese speaker. In this case the man in the Chinese Room manipulates Chinese symbols mechanically without understanding what they mean. Yet his answers are indistinguishable from those of the native Chinese speakers.

The above situation shows that a computer has syntax, but not semantics. Indeed, understanding a language having mental states at all, involves more than just having a bunch of formal symbols. It involves having a meaning attached to those symbols. And a digital computer, as defined above, cannot have more than just formal symbols, because it operates, as Searle says, in terms of its ability to implement programs. As these programs are purely formal they cannot have semantic content.

The supporters of AI argue that we can feed the understanding of Chinese into a robot. If the robot operates the Chinese symbols property, would not that be enough to guarantee that it understands Chinese? Searle replies that the robot lacks understanding. Even though it might behave exactly as if it understands Chinese, it would still have no way of getting from the syntax to the semantics of Chinese. Thus there is no way that the supporter of strong AI can argue that the mind consists of purely formal or syntactic operation, and the mind is nothing but a computing machine.

Searle's Chinese Room Argument is concerned with the issue of understanding and the question whether an appropriately sophisticated computer action can be said to have mental properties. It is concerned with some programs that purport to simulate human understanding by providing replies to questions in Chinese by following purely formal rules. However, despite the appearance of understanding that is involved in the computational output when it performs the computations, no such understanding is actually experienced by the computer performing manipulations that enact these computations. Accordingly, Searle argues that the mental quality of understanding cannot be just a computational matter. It is because the computer is unable to duplicate human intelligence, though it has the ability to simulate the latter. Here, the key distinction is between duplication and simulation. And no simulation by itself ever constitutes duplication. At the end of the argument, he says, "for any artefact that we might build which had mental states equivalent to human mental states, the implementation of a computer program would not by itself be sufficient. Rather the artefact would have to have powers equivalent to the power of the human brain."

In his paper "*Computing Machinery and Intelligence*," Turing has suggested an operational test for the machine-intelligence in the form of an 'imitation game'. Accordingly, if a computing machine can give responses to questions that make it impossible for us to distinguish this computer from a fellow human being, then we can test whether a machine can think or not.

Searle objects to the Turing Test on the ground that the normal criteria we apply in ascribing intelligence to persons are based on behavioural, biological, and phonological evidence. According to him, normal human beings have intentionality, consciousness, and free will, etc., which the machines like computers lack. In effect, the Turing test is a form of reverse discrimination against humans, as it shows the humans in a poorer light in comparison the machines which are made by the human beings.

Searle is certainly correct in saying that merely instantiating a computer program is not sufficient for the possession of our kind of mentality. Mere exhibition of a formally accurate operation does not suffice to make the operation intelligent in the human sense. The fact that human beings have intelligent operations of mind is biologically conditioned and cannot be transformed to non-human machines.

Searle offers two different sets of criteria for applying the expression 'intelligent behaviour.' One of these sets consists of third-person or 'objective' criteria that are not necessarily of any subjective psychological interest whatever. But the other set of criteria are essentially subjective and involve the first-person points of view. According to him, 'intelligent behaviour' on the second set of criteria involves thinking, and thinking is essentially a subjective mental process. Now, if we adopt exclusively the third-person criteria to intelligent-behaviour, then computers, such as not to mention pocket calculators, cars, thermostats, and indeed just everything in the world, engage in intelligent behaviour. But this yields no specific result regarding intelligent behaviour of machines.

## References


- Nikolić, Danko (2015), "Practopoiesis: Or how life fosters a mind", Journal of Theoretical Biology, 373: 40–61, doi:10.1016/j.jtbi.2015.03.003, PMID 25791287
- Chalmers, David J (1996), The Conscious Mind: In Search of a Fundamental Theory, Oxford University Press, New York, ISBN 0-19-511789-1
- Searle, John (1980), "Minds, Brains and Programs" (PDF), Behavioral and Brain Sciences, 3 (3): 417–457, doi:10.1017/S0140525X00005756
- Brooks, Rodney (1990), "Elephants Don't Play Chess" (PDF), Robotics and Autonomous Systems, 6: 3–15, doi:10.1016/S0921-8890(05)80025-9, retrieved 2007-08-30
- Gillett, C. (2007). "A Mechanist Manifesto for the Philosophy of Mind: The Third Way for Functionalists". Journal of Philosophical Research, invited symposium on "Mechanisms in the Philosophy of Mind", vol.32, pp. 21-42.
- Russell, Stuart J.; Norvig, Peter (2003), Artificial Intelligence: A Modern Approach (2nd ed.), Upper Saddle River, New Jersey: Prentice Hall, ISBN 0-13-790395-2
- Turing, Alan (October 1950), "Computing Machinery and Intelligence", Mind, LIX (236): 433–460, doi:10.1093/mind/LIX.236.433, ISSN 0026-4423, retrieved 2008-08-18
- Piccinini G (2010). "The mind as neural software? Understanding functionalism, computationalism, and computational functionalism". Philosophy and Phenomenological Research. 81 (2): 269–311. doi:10.1111/j.1933-1592.2010.00356.x
- Penrose, Roger (1989), The Emperor's New Mind: Concerning Computers, Minds, and The Laws of Physics, Oxford University Press, ISBN 0-14-014534-6
- Emil Post (1936), "Finite Combinatory Processes—Formulation 1", Journal of Symbolic Logic, 1, 103–105, 1936. Reprinted in The Undecidable, pp. 289ff
- Chalmers, David (30 March 1996), The Conscious Mind: In Search of a Fundamental Theory, Oxford University Press, ISBN 978-0-19-983935-3
- Turing, Alan (October 1950), "Computing Machinery and Intelligence", Mind, LIX (236): 433–460, doi:10.1093/mind/LIX.236.433, ISSN 0026-4423, retrieved 2008-08-18

- Piccinini G., Craver C. F. (2011). "Integrating psychology and neuroscience: Functional analyses as mechanism sketches". Synthese. 183 (3): 283–311. doi:10.1007/s11229-011-9898-4
- Yee, Richard (1993), "Turing Machines And Semantic Symbol Processing: Why Real Computers Don't Mind Chinese Emperors" (PDF), Lyceum [1], 5 (1): 37–59
- McGinn, Collin (2000), The Mysterious Flame: Conscious Minds In A Material World, Basic Books, p. 194, ISBN 0786725168
- Hearst, Marti A.; Hirsh, Haym; Bundy, A.; Berliner, H.; Feigenbaum, E.A.; Buchanan, B.G.; Selfridge, O.; Michie, D.; Nilsson, N. (Jan./Feb. 2000), "AI's Greatest Trends and Controversies", IEEE Intelligent Systems, 15 (1): 8–17, doi:10.1109/5254.820322

# 4

# Non-Computational Theories of Mind

The non-computational theories studied under this chapter are theory of mind and Cartesian theory. These theories focus on the ability of the mind to assign mental states to other persons. This chapter discusses non-computational theories in a critical manner providing key analysis to the subject matter.

## Theory of Mind

Emergentism as a theory of the mind is compatible with the non-computational theory of mind. According to emergentism, the higher-level quality emerges form the lower level of existence and has its roots therein, but it emerges there from, and it does not belong to that level, but gives rise to a new order of existence with special laws. Whereas Samuel Alexander says, "the higher-level of quality emerges from the lower level of existence and have its roots therein, but it emerges there from, and it does not belong to that lower level, but constitutes its possessor a new order of existence with its special laws of behaviour. The existence of emergent qualities thus described is something to be noted as some would say, under the compulsion of brute empirical fact, or, as I should prefer to say in less harsh terms, to be accepted with the 'natural piety' of the investigator. It admits no explanation." Here, emergence refers to the fact that in the course of evolution new things and events occur, with unexpected and unpredictable properties. Things and events are new in the sense in which a great work of art may be described as new. Every genuine emergence introduces novelty into the world. To say that an emergent characteristic is a novel means, firstly, it is not simply a rearrangement of pre-existing elements, although such rearrangement may be one of its determining conditions. Secondly, the characteristic is qualitatively, not just quantitatively, unlike anything that existed before in history. Thirdly, it is unpredictable not only on the basis of knowledge available prior to its emergence but even on the basis of ideally complete knowledge of the state prior to its emergence. These points permit a distinction to be made between what is new in the sense of being a fresh combination of old factors and what is novel in the sense of being qualitatively unique and unpredictable.

We will find out how consciousness emerges from material properties, and how the emergent property of consciousness cannot be explained in a functional or computational way. Some philosophers argue that consciousness might be an emergent property, in a sense that it is still compatible with materialism. It is also often held that emergent properties are unpredictable from low-level properties. However, it can be argued that these properties are new in an ontological sense. What is interesting about these properties is that they are not obvious consequences of the low-level properties. But they are still causally supervenient on low-level facts.

Following the above argument, we can put argue that the phenomenon of consciousness could rise only in the presence of some non-computational physical processes taking place in the brain. As we know, living human brains are ultimately composed of the same material satisfying the same physical laws as are the inanimate objects of the universe. There is the Cartesian view that consciousness arises only in humans, and that animals are inanimate automata; a view which is clearly pre-evolutionary. And we have reason to accept the view that there are lower and higher states of consciousness. Moreover, the most reasonable view seems to be that consciousness is an emergent property of animals arising under the pressure of natural selection. If this is so, then the question is; how does consciousness arise from to antecedent conditions in the physical universe? This question is still unanswered.

Even if the observation of the behaviour of the amoeba created the strong impression that it is conscious. And we can find symptoms of activity and initiative in its behaviour. That activity and behaviour is something different from what happens in the neurons. The human brain is estimated to have ten thousand millions of neurons, and there are also thousands of synaptic relations among neurons. But, the qualities, which exist in consciousness, are not found in the neuronal relations. There is a new emergent entity in consciousness which did not exist in the neurons. Because, the emergent properties of consciousness are ontologically new. The problem of emergence in this context starts with 'life' and it should be remembered that 'brain' is not just a pice of inanimate matter, but a part of the 'living' body. As Daya Krishna remarks, "it is not even clear whether those who want to deny the 'reality' of consciousness want to deny the reality of 'life' also. The 'body' they talk about is a 'living body', the 'brain' they are fond of is the brain that is 'alive'. Take 'life' away and everything 'dies', 'ceases', at least as we 'live' it and 'feel' and 'know' it."[2] Here, Daya Krishna is trying identify 'life' with 'consciousnesses'. Here, like consciousness, life also emerges from human body and identify it (life) with consciousness. In fact, the problem of emergence of 'life' is a far wider one but here we are concerned with human 'life' only.

Now, the question is; why is it that the phenomenon of consciousness appears to occur, as far as we know, only in 'living' beings, although we should not rule out the possibility that consciousness might be present also in other appropriate physical systems? The second question is, how could it be that such a seemingly ingredient as non-computational behavior presumed to be inherent in the actions of all material things, so far has entirely escaped the notice of physicists? The first question is that the question is related with the subtle and complex organization of the brain, but that alone could not provide a sufficient explanation. Penrose clearly writes, "I am contending that the faculty of human understanding lies beyond any computational scheme whatever. If it is microtubules that control the activity of the brain, then there must be something within the action of microtubules that is different from mere computation."[3] Here, he says that this inanimate matter is microtubules that control the activity of the brain because there is life in it (brain). The action of microtubules is different from mere computation. He point out that such actions are non-computational actions, in which 'life' is related to consciousness.

The above statement leads to the question, is there any evidence that the phenomenon of consciousness is related to the action of microtubules in particular? It must also be the case that the detailed neural organization of the brain is fundamentally involved in governing what form that consciousness must take. For Penrose, if that organization were not important, then our livers would evoke as much consciousness as do our brains. He puts it, "What the preceding arguments strongly suggest is that it is not just the neural organization of our brains that is important. The cytoskeletal underpinning of those very neurons seems to be essential for consciousness to be present." But it's not a cytoskeleton as such that is relevant, but some essential physical action that biology has so cleverly contrived to incorporate into the activity of its microtubules. Moreover, it may be pointed out that in our brain there is an enormous organization, and since consciousness appears to be a very global feature of our thinking, it seems that we must look to some kind of coherence on a much larger than that level of single microtubules or even single cytoskeletons. And there is some kind of useful non-computable action involved, which Penrose takes to be an essential part of consciousness. Secondly, we must expect that vestiges of such non-computability should also be present, at some indiscernible level, in inanimate matter. But yet the physics of ordinary matter seems to allow no room for such non-computable behaviour.

Jaegwon Kim, in his article on 'Supervenience' argued that there is a striking similarity between emergence and supervenience. According to Kim, "higher-level properties notably consciousness and other mental properties, emerge when, and only when, an appropriate set of lower-level (basal conditions) are present, and this means that the occurrence of the higher properties is determined by, and dependent on, the instantiation of appropriate lower-level properties and relations. In spite of this, emergent properties were held to be 'genuinely novel' characteristically irreducible to the lower level processes from which they emerge." Then, the concept of emergence combines the three components of supervenience delineated above, namely, property covariance, dependence, and non-reducibility. Thus, emergentism can be regarded as the first systematic formulation of non-reductive physicalism.

This thesis makes the mental life supervenient on its physical background. That is to say, according to this thesis, the mental states are not reducible to but are supervenient on the physical states in such a way that whatever changes take place in the physical states must make a difference to the mental states well. No two things could differ in a mental respect unless they differed in some physical respect, that is, imperceptibility with respect to physical properties entails indiscernibliity with respect to mental properties. That is the core idea of mind-body supervenience.

Thus supervenience understood in the strong sense makes room for a nomological dependence of the mental on the physical such that the physical states are necessarily responsible for the mental states. As Kim points out, one must notice that the mental is dependent on the physical but not vice versa, because the mental states are directly a consequence of the physical states. The mental states themselves do not determine the physical states. In that sense the mental states remain nomologically dependent on the physical universe according to thesis of strong supervenience.

Kim, in his article on *'The Non-Reductionist's Troubles with Mental Causation'* mentions that the non-reductive physicalism consists of following theses.

i. All concrete particulars are physical.

ii. Mental properties are not reducible to physical properties.

iii. All mental properties are physically realized; that is, whenever an organism, or system, instances a mental property M, there is a physical property P such that P realizes M in organisms of its kind.

iv. Mental properties are real properties of objects and events; they are not merely useful aids in making predications or fictitious manners of speech.

Therefore, we find that these four basic tenets bring non-reductive physicalism very close to 'emergentism'. In fact, non-reductive physicalism of this variety is best viewed as a form of emergentism. Emergentists in general accepted purely materialist ontology of concrete physical objects and events. For example, Kim following, Samuel Alexander, one of the principal theoreticians of emergence school, argues that there are mental events over and above neural processes: Alexander says, "We thus become aware, partly by experience, partly by reflection, that process with the distinctive quality of mind or consciousness is in the same place and time with a neural process, that is, with a highly differentiated and complex process of our living body. We are forced, therefore, to go beyond the mere correlation of the mental with these neural processes and to identify them. There is but one process which, being of a specific complexity, has the quality of consciousness. It has then to be accepted as an empirical fact that a neural process of a certain level of development processes the quality of consciousness and is thereby mental processes; and, alternatively a mental processes is also a vital one of a certain order."

However, the emergentist doctrine that emergent properties are irreducible to the physical conditions out of which they emerge is familiar; this irreducibility claim is constitutive of the emergentist metaphysical world–view. Although the emergentists' idea of reduction or reductive explanation diverges from the model of reduction implicit in current anti-reductionists argument, the philosophical significance of the denial of reducibility between two property levels is the same. The higher-level properties, being irreducible are genuinely new addition to ontology of the world. For example, Samuel

Alexander says: "Out of certain physiological conditions nature has framed a new quality mind, which is therefore not itself physiological though it lives and moves and his its being in physiological conditions. Hence it is that there can be and is an independent science of psychology... No physiological constellation explains for us why it should be mind."

The strong supervenience thesis does not bridge the gap between mental and physical because it fails to account for how the mental states with their qualitative content arise at all in a material environment. The gap between the physical and mental remains wide because it is not known how the mental world can be explained. Now the question is: Is it not possible that the mental life not be there even if the physical universe exists perfectly? That is to say, there are possible worlds in which all the physical states of the present universe are there but there is no conscious state at all. For example, robots behave like human beings but lack consciousness. The behaviour itself is not consciousness. And if consciousness is the same in all organisms like material things, then there will be no qualitative difference between the human and non-human. Therefore, we cannot prove that consciousness is supervenient on the physical world.

John R. Searle has given an example, which will make the above thesis more legitimate. Suppose, we have a system S, and the elements of system are A, B, C.... S might be a stone and the elements might be molecules. There will be features of S that are not, or not necessarily features of A, B, C, . . . But there are some features which are 'causally emergent system features. Solidity, liquidity and transparency are examples of causally emergent system features. In this connection, we should remind us that 'life' is an 'emergent' property and if there were no 'life', there would be no consciousness either. For example; as we know water is the combination of two-hydrogen and one-oxygen. But there are qualitative difference between water on the one hand, and the hydrogen and oxygen on the other hand. The qualities which we will find in the water, will not find in the oxygen and hydrogen. In the same way, there is difference between consciousness and matter because there is a qualitative difference between the two. The qualities which emerges from consciousness will not possible to explain in the mechanical/functional way, but it needs separate explanation, and its explanation is non-reductive explanation, that is, self-explanation.

The above definition shows that consciousness is a causally emergent property of systems. It is an emergent feature of creation systems of neurons in the same way that solidity and liquidity are emergent features of the system of molecules. Thus, the existence of consciousness can be explained by the causal interactions between elements of the brain at the micro level, but consciousness cannot itself be deduced or calculated from the sheer physical structure of the neurons without some additional account of the causal relations between them.

Now the question is: Why is consciousness an irreducible feature of physical reality? There is a standard argument to show that consciousness is not reducible in the way that material qualities are. For example, I am now in a certain conscious state such as pain. Now the question is: what fact in the world corresponds to my statement. 'I am now in pain'? Here is the fact that I have now certain unpleasant conscious sensations, and I am experiencing these sensations from my experience. It is these sensations that are constitutive of my present pain. But the pain is also caused by certain underlying neurophysiological processes consisting in large part of patterns of neuron firing in my brain. If we reduce the first-person sensation of pain to the third-person patterns of neurons firing, then we try to say that the pain is really 'nothing but' the patterns of neuron firings. If this is so, then we are leaving the essential features of pain. No description of the third-person type would convey the first-person character of pain because the first-person features are different from the third-person features. Nagel states this point by contrasting the objectivity of the third-person features with the what-it-is-like features of the subjective states of consciousness. As Nagel point out, "Conscious experience is a widespread phenomenon. It occurs at many levels of animal life, though we cannot be sure of its presence in the simpler organisms, and it is very difficult to say in general what provides evidence of it... no matter how the form may vary, the fact that an organism has conscious experience *at all* means, basically, that there is something it is like to *be* that organism...But fundamentally an organism has conscious mental states if and only if there is something it is like to *be* that organism—something it is like *for* the organism." Thus I know that I am in pain is a different sort of knowledge than my knowledge that you are in pain.

The feeling of pain indicates that there is close relation between consciousness and self-consciousness. This is due to the emergence of self-consciousness out of consciousness and thus making it radical different from what is it, if it is at all human levels. But Daya Krishna remarks that the development of robotics denies the reality of consciousness because of this 'self-consciousness' knowledge have 'self-determination' and deny the existence of mind or consciousness. This 'self-consciousness' has forgotten its dimension of 'knowing', 'feeling', and 'willing', the last resulting in the transformations through technology that has obsessed the modern mind to such an extent that it has gone to the extent of denying its own reality and considering the 'matter' alone as 'real'. But yet, matter, though resistant, is flexible, agreeable to change which consciousness does not seem to be in the same sense. The denial of 'I-consciousness' which is an inevitable accompaniment of self-consciousness. Thus Daya Krishna remarks, "The 'real' causal role of consciousness, however, becomes clear when self-consciousness comes into its own and discovers that it can directly affect consciousness, and indirectly everything else through imaging, intending, thinking, attending, concentrating, reasoning and the other myriad activities which man has encountered in himself and developed through a long process . . .."

Therefore, from the above exploration, it follows that once consciousness emerges from physical properties, it will never be reduced to it. This shows that emergentism cannot support functionalism or computationalism. Because a functionalist explains consciousness or mind in the reductive way, whereas an emergentist explains consciousness in a non-reductive way. Consciousness makes the mind-body problem really intractable. The reductionist shows that the mind-body problem is not a real problem. For him or her, there is no explanatory gap between mind and body. We have to find out why these arguments do not help us to understand the relation between mind and body. Without consciousness the mind-body problem would be less interesting, but with consciousness it seems hopeless.

The hard problem of consciousness is the problem of experience, especially to first-person character which cannot be explained within a scientific framework. Cognitive science can explain a system's functions in terms of its internal mechanism. But it is not possible to explain what it is to have subjective experiences, because it is not a problem about the performance of functions. As Nagel argues, "Conscious experience is wide spread phenomenon . . . fundamentally an organism has conscious mental states if and only if there is something that it is like to be that organism –something it is like for the organism." In recent times, all sorts of mental phenomena have yielded to scientific explanation, but consciousness has stubbornly resisted this explanation. Many philosophers and scientists have tried to explain it, but the explanations always seem to fall short of the target. Now the question is: why is it so difficult to explain? According to Chalmers, cognitive science has not explained, why there is conscious experience at all. When we think and perceive, there is a whir of information processing, but there are also subjective individual aspects of consciousness, which go beyond the information processing.

Chalmers writes, "When it comes to conscious experience, this sort of explanation fails. What makes the hard problem hard and almost unique is that it goes beyond problems about the performance of functions. To see this, not that even when we have explained the performance of all the cognitive and behavioural functions, in the vicinity of experience –perceptual discrimination, categorization, internal access, verbal report –there may still remain a further question: Why is the performance of these functions accompanied by experience?" According to him, even if all the functions of a system are well articulated, there is further question as to why there is any experience at all accompanying their function. Cognitive science fails to explain why there is any experience at all, even though it explains all the brain functions.

According to Chalmers, the hard problem of consciousness consists in the 'why' questions regarding consciousness. But the question is: why is the 'hard' problem so hard? And why are the easy problems so easy? The easy problems are easy because they concern the explanation of cognitive abilities and functions. To explain a cognitive function, we need a mechanism that can perform the function. The cognitive sciences offer this type of explanation and so are well suited to the easy problem of consciousness. On the other hand, the 'hard' problem is 'hard', because it is not a problem about the performance of functions. The problem persists even when the performance of all the relevant functions are explained. Chalmers says, "I suggest that a theory of consciousness should take experience as fundamental. We know that a theory of consciousness requires the addition of something fundamental to our ontology, as everything in physical theory is compatible with the absence of consciousness. We might add some entirely new non-physical feature, from which experience can be derived, but it is 'hard' to see what such a feature would be like. More likely, we will take experience itself as a fundamental feature of the world, alongside mass, charge, and space-time. If we take experience as fundamental, then we can go about the business of constructing a theory of experience."

The reductionists have not solved the 'hard' problem of consciousness because, as we have seen, it has explained consciousness only in terms of the 'easy' problem of consciousness. Easy problems are all concerned with how a cognitive or behavioural function is performed. These are questions about how the brain carries out the cognitive task, that is, how it discriminates stimuli, integrates information and so on. Whereas the hard problem of consciousness goes beyond the problems about how functions are performed. If scientific view of mind tries to give a definite definition of consciousness then it leaves out the explanatory gap, that is, there is no explanatory gap between mind and body. Because there is no distinction between mind and body. Mind can be explained in terms of body, and there is nothing called the mind, since the mind itself is a part of the body. If this is so, then it leaves out subjective experience, and opts for the third-person perspective of consciousness.

Consciousness makes the mind-body problem really intractable. The reductionists deny that there is a mind-body problem at all. For them, there is no explanatory gap between mind and body. Because there is no distinction between mind and body. Mind can be explained in terms of body, and there is nothing called the mind, since the mind itself is a part of the body. Therefore, for them, the mind is reductively explainable in terms of body. On the other hand, many philosophers hold that mental states are not reducible to any physical state(s). That is, the mental states are not reductively explainable. Chalmers argues that no reductive explanation of consciousness can succeed, because there is subjective quality of experience. Therefore, he argues that this quality of consciousness makes it different from all other properties, including emergent biological properties such as life.

The essence of body is spatial extension, the essence of mind is thought. Thought is taken to be the defining attributes of mind which is an incorporeal substance a substance that is non-spatial in nature. Descatres writes, "By the term 'thought, I understand everything which we are aware of as happening within us, in so far as we have awareness of it." What follows from Descartes' view is that consciousness is essentially a first person, subjective phenomena, and conscious states cannot be reduced or eliminated into third-person. Therefore, it is consciousness, which makes the explanatory gap between the first person and third-person perspective. According to the Cartesian concept of mind, we have access to the contents of our own minds in a way denied to us in respect to matter. There is something special about our own knowledge of our own minds that naturally goes with the Cartesian view.

Pradhan argues that the mental life with its qualia cannot be nomologically determined by the physical conditions of the universe. The following are the reasons for the thesis that the mental life is independent of the physical body, though they co-exist: "(A). The qualia of the mental states cannot be reproduced in an artificial machine like a robot or a machines table; they are unique to the person concerned. (B). The qualia are the essence of consciousness and so must be intrinsic to the conscious subjects." Thus Pradhan concludes that the intelligibility gap between the qualia and the physical world remains, as the qualia are understood widely as belonging to the conscious subjects.

Consciousness makes the gap between mind and body, and 'subjectivity' is its most troublesome feature. Self is the subject, which encompasses our feelings, thinking, and perception. The qualitative character of experience is what it is like for it's subject to have the experience. As Nagel puts it, "Conscious experience is a widespread phenomenon. It occurs at many levels of animal life, though we cannot be sure of its presence in the simpler organisms, and it is very difficult to say in general what provides evidence of it... no matter how the form may vary, the fact that an organism has conscious experience *at all* means, basically, that there is something it is like to *be* that organism...But fundamentally an organism has conscious mental states if and only if there is something it is like to *be* that organism—something it is like *for* the organism."

As we have seen, subjectivity cannot be explained reductively. Again, as Nagel argues, "It is not analyzable in terms of any explanatory system of functional states, or intentional states, since they could be ascribed to robots or automata that behaved like people though they experienced nothing." There is a subjective feeling attached to our conscious experience because subjective feelings are the outcome of our conscious experience. That is, consciousness itself cannot be established simply on the basis of what we observe about the brain and its physical effects. We cannot explain which property of the brain accounts for consciousness. Distinct cognitive properties, namely perception and introspection, necessarily mediate our relationships with the brain and with consciousness. We cannot understand how the subjective aspects of experience depend upon the brain that is really the problem.

Consciousness is essentially subjective because this is not a mechanical state, as many philosophers believe but it is 'I' or 'subject' who experience. Some of these biological systems are conscious and that consciousness is essentially subjective. The term 'pain' is subjective as it is not accessible to any observer, because it is a first person experience. For example, I have a pain in my leg. In this case, the statement is completely subjective. The pain itself has a subjective mode of existence. As Searle puts it, "Conscious states exist only when they are experienced by some human or animal subject. In that sense, they are essentially subjective. I used to treat subjectivity and qualitativeness as distinct features, but it now seems to me that properly understood, qualitativeness implies subjectivity, because in order for there to be a qualitative feel to some event, there must be some subject that experiences the event. No subjectivity, no experience."

That is to say that the qualitative experience can exist only as experienced by some subjects. Because conscious states are subjective in this sense, it is legitimate to hold that there is first-person ontology, as opposed to the third-person ontology of mountains and molecules, which can exist even when there are no living creatures. Therefore, subjective conscious states have first-person ontology because they exist only when they are experienced by a subject as self. It is 'I' who has experience and in this sense, it has the subjective existence. This gap between the self and the body not only establishes explanatory gap, but also gives the ontology of first-person. Therefore, the 'subjectivity or 'I' is the central problem of the explanatory gap. Cognitive science tries to explain how conscious experience arises from the electrical process of the brain. But it cannot show how and why conscious states belong to the 'subject' or 'I'. This qualitative feature of mental states brings the existence of qualia, which are the qualitative experiences of the human mind.

Searle has been a great critique of supervenience and emergence theory of mind. The Searlean account of *mentalcausation* does refer to building up link between the mental states and the brain states showing the *downwardmovement*. However, this link is further intentionalized with irreducible mental features like intentionality, subjectivity, freedom, etc. This irreducibility thesis maintains the notion of emergence of the mental states but does not find the notion of *supervenience* as a sound theorization of the mind. Biological naturalism advocates for non reductionism and critiques the realizability thesis argued by the emergentism advocated by J. Kim.

John Searle also develops the idea of *causal closure* while advocating *biologicalnaturalism*. The biological naturalism as a theory of mind holds the view of emergentism. The mind has evolved biologically along with the human body constituting of atoms and molecules. This biological creation is a part of the natural world. The human body is governed by the natural laws. The natural laws draw the *causal closure* unfolding the naturalistic project of Searle's theorization of the mind. Moreover, human as an embodied being requires special explanation with regard to the emergence of the mental phenomena. In fact, there are two significant aspects to the Searlean theorization of the mind conceived within the naturalistic paradigm. They are the notion of *emergence* and the notion of *irreducibility*. For Searle, consciousness or the mental phenomena like intentionality, subjective feeling, freedom, etc., are the emerging properties. As emergent properties they have *appeared* out of the causal function of the brain processes. The causal functions sustains biologically between the brain processes and the mental processes. Searle hypothesizes this causal process in his *biological naturalism*. The hypothesis advocates, "Mental states are *caused by* the brain processes and *realized in* the brain processes."

The brain is the locus of causing and realizing the mental states. In other words, the notion of *causing* mental states and *realizing* them are happening at the realm of the physical. In that domain mental states have their own intrinsic features like intentionality, phenomenality, subjectivity, etc. These mental features remain distinct and irreducible to the mere neurophysiology and neurochemistry of the brain. The brain being the constitutive patterns of these layers of functions, cause the emergence of the mental and prevails on *causing* the behaviour. This causal emergence of the mental, Searle illustrates, "Consciousness is causally emergent property of systems. It is an emergent feature of certain system of neurons in the same way solidity and liquidity are emergent features of system molecules. The existence of consciousness can be explained by the causal interaction between elements of the brain at the micro level, but consciousness cannot itself be deduced or calculated from the sheer physical structure of the neurons without some additional account of causal relations between them." The brain being the micro level constituent of consciousness is different from the emergent mental states. The mental sates are conscious and belong to the macro-level. This micro-macro stratification requires different types of explanation though they functionally operate in the realm of the physical. As the macro level function does affect the behaviours of the person, the micro level does cause them and involve in realizing them.

The involvement of the physical is inevitable as a constituting cause of the mental. But this constituting cause does not suffice fully to explain the intrinsic features of the mental states and their interaction. Explanation of their interaction requires an additional account of causal relations. Moreover, that may not entirely put off his thesis of irreducibility. For him, 'consciousness can be caused out of the brain processes but it would have certain 'causal power that cannot be explained by' the micro level system features of the brain. This thesis would entail another notion of emergence, that is, "... consciousness could cause things that could not be explained by the causal behaviour of the neurons. The naïve idea here is the consciousness gets *squirtedout* by the behaviour of the neurons in the brain, but once it has been squirted out, it then has a life of its own." Consciousness, in this respect, as an emerged system feature is causally efficacious to build a system and determine the function of the mental functions entirely. In other words, this notion of emergence would imply the autonomy of the mental that could be maintained out of the functions of the brain processes. Searle's naturalistic schema does not approve it.

However, he intends to argue that the notion of irreducibility of the mental stating that "Consciousness is causally emergent property of the behaviour of neurons and so consciousness is causally reducible to the brain processes. But – and this is what seems shocking – a perfect science of the brain would still not lead to an ontological reduction of consciousness in the way our present science can reduce heat, solidity, colour and sound." Searle lays emphasis on non-reductionistic theory of mind showing that reductionism ultimately advocates ontological reduction. The ontological reduction of the mind to the brain processes eliminates the reality of the mental phenomena or consciousness. The reality of consciousness lies in the first person subjective experiences and feelings. This reality appears in the macro level. In other words, the *appearance* of consciousness in our feelings, experiences and actions is not only different from the physico-chemical processes of the neurons of the brain but also shows that the appearance is reality.

However, the sustained causal interaction between the brain and the mind shows the upward and the downward movement of the causal functions respectively. As the causal power of the brain is causing consciousness, similarly the *causal power of the mind* is responsible for voluntary movement/ action, feelings, experiences, etc. The causal power of the mind shows a downward mobility causing/ *transforming* mental states into action. The existence of the mental hence supervenes on the physical states and processes of the brain. Searle writes, "On this characterization of the supervenience relation, the supervenience of the mental one the physical is marked by the fact that physical states are causally necessary for the corresponding mental states.... this sort of supervenience is a causal supervenience." The causal supervenience claims that the mental interacts with the physical.

Interestingly Kim denies the causal interaction between the mental and the physical. The mental being supervenient on the physical cannot have causal impact on the physical. Rather mental properties are *causally impotent*. The causal impotency does not imply the reducibility. Firstly, it is because there is no causal link between supervening features of the mental and its basal level conditions. The supervenient relation only expresses a *vertical* dependency relation. But this expression has nothing to do with the real *horizontal* causal relations that the base level holds while causing another mental/ brain state. (e.g. P *causingP**) While arguing for supervenience, Kim understands that 'it is not possible to have causation in the mental realm without causation being crossing into the physical realm.' Its significance lies in a state of relation where 'we can no longer isolate the causal relation at higher level and its bottom level which would entail cross level causal relation'. Neither Kim would talk about the cross level causal relationship between the physical and the mental nor would he talk about the level bound autonomy, which would be 'inconsistent with the notion of supervenience.' If his notion of supervenience is acceptable then it does not fit with the notion of emergence. The notion of emergence shows the *gap* between two levels of complex functions. Each level function will have causal impact, i.e. the power of making a complex structure and further giving rise to certain emergent properties. This process may not have direct causal link but one getting emerged out of the other and being able to sustain a complex function itself causally empowers the emergent properties. This is very much a part of the Searlean theorization of the intentionality emerging as intentional mental feature out the brain functions. Intentionality of the mental states builds up the network of mental states just being empowered with certain causal power. This intentionality is not thrown away from its material ontological basis. The gap between the causal process of the physical level and the causal process of the intentional-mental level is a conceptual gap, which any emergence theory would advocate. Kim in this regard is an exception to this idea. In Ned Block's terminology the causal power of the emerged mental property *drains away* making the mental supervenient.

On the other hand, if supervenient causal relationship rests on the horizontal position then there is no *level* of emergence *per se*. Assume that one level of emergence is the level of the mental; it would only constitute a parallel mental level as supervenient level. At this level, we would have the emergence of discrete mental states corresponding to change and occurrence of the brain states at physical basal level. The emerging mental states would be discrete in the sense that they would not have any *unifying* properties to relate with. Even if we maintain the condition of simultaneity and multiple realization conditions for emergent states and their realization, still lack of unifying character to build them would imply the loss of *subjectivity*. Therefore, the causal chain that builds in is only at the level of physical which excludes the mental. Kim writes, "Note that *P*'s causation of *P** cannot be thought of as a causal chain with M as an intermediate causal link; one reason is that the *P* – to – *M* relation is not a causal relation. Note also since that *M* supervenes on *P*, *M* and *P* occur at the same time." This notion of *exclusion* of the mental not only eliminates the *feeling* of *subjectivity* but also eliminates the notion of *level per se*. That is, drawing up differences in the levels with horizontal causal mobility may imply emergence but would not imply levels. How does one talk about vertical level of emergence with reference to the formation of water? The two levels of existence are causally efficacious and functional so far as their *emergence* and *submergence* are concerned.

The notion *simultaneity* needs to be discussed with regard to the notion of *overdetermination*. Kim does not want to defend the notion of *overdetermination* because it involves the causal power of the mental states along with the physical state for determining the condition of *realization*. Mental properties are *extrinsic* and *relational*. They are extrinsic because 'no mental property would have non-physical realizations. Rather they must be physically realized.' Mental properties are 'second-order functional properties' relative to the 'first-order physical functional properties.' The former do not constitute the functional properties; rather it shares nomological relation with the latter. If mental properties intervene in relationship with the physical for determining condition of realization then it would imply that the physical is supervening on the mental. There is no such supervenience justified because it would entail reductionism. Thus, for Kim, overdetermination entails reductionism. He finds that the Searlean mental realism advocates overdetermination. It is because the mind intentionally or *consciously* acts *in* the world. The intentional mode of relation is effectively felt and shows the realization of the mental states. The notion of overdetermination with reference to *experience* and *realization*, shows *fusion* of top-down and bottom-up causation. This fusion represents the notion of simultaneity: Say, 'I am feeling pain'. It means I have *sensory qualia* of pain, which simultaneously *go* with the *intentional qualia*. The latter implies a *dynamic* intentional representation of *feeling* of pain, whereas the former represents a *stable* sensory representation of pain. For Searle, both are two different modalities of processing the representation. Moreover, the feeling of pain needs intentional representation to get relived from it. So, 'how do I make this representation?' is certainly an important point of referring to thoughts and its dynamic relations, which helps in forming a synthesis/ unity of those representational states. It is obvious that the process of synthesizing would have physical basis. The time involved in the physical simulation (whatever mille seconds) is unlikely to be at the same time involved for intentional representation. The simultaneity *reduces* the differences in the processes. If simultaneity zeroes down the levels of processing and levels of interferences between the intentional and the sensory activities then how is that Kim is missing the 'homogeneous natural relations' between the physical and the mental. In this regard, supervenience would entail reductionism.

The Searlean thesis does not imply overdetermination rather the functional correlation, implying the functional autonomy. The notion of functional correlation between the top-down mental causation relating to the bottom-up causal processes is a conjecture. We are yet to *know* their level of meeting points, and Searle candidly admits that. Where do the physical processes end with emergent mental properties and where do the mental properties reside with the physical, is not only a matter of scientific investigation but also an issue to be delved through philosophical inquiry. Even if we empirically find their point of emergence and point of correlations, still the significance of the mental will not be nullified; it would have its own level of description. The Searlean notion of levels of description is therefore significant, but does it not divide the reality falling into the track of property dualism?

However, the notion of property dualism could not be strengthened due to Searle's disclaimer for *ontological emergence*. Searle, like Kim accepts the 'modest kind of emergence.' Advocating the modest notion of emergentism, he may succeed in showing that ontology of intentionality or the mental could be traced from the ontology of the physical. Moreover, in Searle's interpretation, the mental has the power (causal power) to go beyond the *causal closure* held by the physicalists. Searle is neither physicalist, nor is he a radical naturalist. His naturalism does not advocate reductionism under the banner of unification of science. Nevertheless, he does not disown the laws of the nature operating in the universe and having impact on the body. But, at the same time, we are involved in performing conscious activities, relating ourselves with the world, forming norms and defining various types of rationalities involved in human social life. All these have epistemic significance and are formed by the 'conscious intentionality.' The notion of *conscious intentionality* becomes a ground for experiencing causal relations in the world. The emergence of intentionality as an intrinsic mental feature fulfills the epistemic necessity of being. These epistemic needs are satisfied with the emergent mental properties. The Searlean thesis argues for a higher order explanation of these activities. Then, is it the case that Searle is advocating 'epistemological emergence'?

Defining the notion of epistemic emergence Silberstein and McGreever writes, "A property of an object or system is empistemically emergent if the property is reducible to or determined by the intrinsic properties of ultimate constituents of the object or systems, while at the same time it is very difficult for us to explain, predict, desire the property on the basis of ultimate constituents. Epistemologically emergent properties are novel only at the level of description." In other words, the epistemic emergence talks ultimately about the reduction of the emerged properties to its basic constituent properties of the system. Following this definition of epistemic emergence, Searle's notion of intentionality, subjectivity, etc. as emerged system features/ properties are emergent properties and hence ultimately reducible to the function of the brain. Though Searle is hopeful about the scientific explanation of the consciousness, and determining its relationship with the brain, still he is not in favour of any kind of reductionism. According to him, human intentionality and subjectivity are irreducible features of consciousness. As Searle primarily puts it, "Consciousness is, by definition, subjective, in the sense that for a conscious state to exist it has to be experienced by some conscious subject. Consciousness in this sense has a first-person ontology in that it only exists from a point of view of human or animal subject, an 'I', who has the conscious experience. Science is not used to dealing with phenomena that have first person ontology. By tradition, science deals with phenomena that are 'objective,' and avoids anything that is 'subjective.'" Hence, for Searle, the relationship between subjectivity, intentionality and consciousness needs to be explained from a different macro level whereas the relationship between the consciousness and the brain is to be viewed from the micro level.

Now the question may arise, are these two levels of functions independent of one another or are they interrelated? These functional explanations are independent of each other because they explain things from two view points: the macro level explanation of consciousness or conscious action is from the *first person point of view*, whereas the micro level explanation of human action is from *third person point of view*. The epistemic objectivity of these explanations differs because the explanation presupposes human subjectivity as an objective fact of the reality. The epistemic objectivity stated from third person point of view shows there are entities which have objective existence. These entities do not depend on someone's subjective opinion or point of view rather they have independent existence in the world. So, these points of view bring in the 'distinction in modes of existence, i.e. the ontological sense of the objective subjective distinction.' Searle makes it clear that this epistemic objectivity of science does not include the ontological subjectivity. In this regard, the notion of independence can be discussed with reference to two different respective frameworks in which the emergent conscious features are viewed, they are: the framework of science and the framework of philosophy. Moreover, if the explanations of these frameworks are interrelated, then Searle is offering a functionalists explanation of the mental phenomena. But, obviously, Searle does not accept the functionalism. Though mental phenomena are caused by neural configurations and realized in the same neurological structure, there is no identity as such between the neurophysiological (micro) level and the (macro) mental level. Thus the mental is not just an emergent property but is something real.

Furthermore, this reality of consciousness then could presuppose the ontological emergence in which the mental features likes subjectivity, intentionality, qualia, etc. are having a unified ontological status. So far as the notion of ontological emergence is concerned, Silberstein and McGreeverstates that "features are neither reducible to nor determined by more basic features. Ontological emergent features are basic features of systems or wholes that possess causal capacities not reducible to any of the intrinsic causal capacities of the parts, nor to any of the (reducible) relation between the parts."- Looking at this notion of irreducibility thesis prevailing as one of the strong points in the framework of Searlean theorization of emegentism, David Papineau remarks during the discussion of the paper whether it would entail the notion *metaphysical supervenience*. According to Papineau, metaphysical supervenience presupposes the mental as an ontological category, prior to the ontology of the physical. The answer to such a position ultimately leads to the traditional philosophical position of the Cartesian dualism. Searle is a great critique of the metaphysical positions of the Cartesians. Rather he writes, "As long as we continue to talk and think as if a the mental and the physical were separate metaphysical realms, the relation of brain to consciousness will forever seem mysterious, and we will not have a satisfactory explanation of the relation of neuron firings to consciousness." In this connection, Searle is not accepting the metaphysical positions of Cartesianism, rather for him all the evolutes are causally connected to the nature as a whole. The mind is part of the nature. But this relationship does not hold to any sort of reducibility, that mind or the conscious experience that comes in the unified form is casually reducible to the physical structural properties and functions. It follows that there is a relationship between the mental and the physical but no reductionism is *ever* possible between them. The mental and the physical belong to the realm of the biological having two different layers of existence.

The human body is a complex biological organism. So far as the mind is concerned it has evolved out of the complexity involved in the evolutionary process in general and the neurophysiolgoical structure of the brain/body in the particular. The emergence of the mind from the neural structure shows that Searle would accept the view that mind is dependent on the brain (the most vital part of the body), because the mental phenomena are caused by the brain. But there is no causal law operating in the realm of the mental. On the other hand, actions are intentionally defined. Intentionality is one of the basic features of the conscious mind. Thus, the relationship between the mental and the physical is one of the asymmetrical dependencies. It shows that the mental is dependent on the physical but not reduced to the physical. In this regard, Searle's philosophical position comes close to Kim's nonreductivephysicalism.

However, about the nonreductive standpoint of the Searlean thesis, one needs to go back to the unified feature of human consciousness or human intentionality *per se*. The experiences are unified intentionally which makes a space for not only asserting the ontological subjectivity but also postulating the notion of self. As we have stated earlier, for Searle, "... all conscious states are ontologically subjective in the sense that it only exists as experienced by one." Subjectivity is intrinsic to experience and presupposed as an objective phenomenon. The disclosure of subjectivity in experience, thought and action shows the postulation of self as an intentional being. This self is not only a thinking being but also a performer of free action, rational choice, decision, etc. Thus the self as the representation of intentional subjectivity involves the activities of thinking as well as action in the form of interaction with the world. The intentional subjectivity is constituted of person's attitude, feeling, experiences, moods, etc.Human thinking is not only intentional, but also normative. Being intentional it shows how the content of thought and expression is experienced and understood.

## Problem of Mental Causation

The problem of mental causation is a conceptual issue in the philosophy of mind. That problem, in short, is how to account for the common-sense idea that intentional thoughts or intentional mental states are causes of intentional actions. The problem divides into several distinct sub-problems, including the problem of causal exclusion, the problem of anomalism, and the problem of externalism. However, the sub-problem which has attracted most attention in the philosophical literature is arguably the exclusion problem.

## Description

The basic problem of mental causation is an intuitive one: on the face of it, it seems that mental events cause physical events (and vice versa), but how can mental events have any causal effect on physical events? Suppose, for example, some person, John, orders dessert after dinner. It seems that at least one cause for such a physical, behavioral event is that John desired to have dessert and believed that by ordering dessert he would be able to soon have dessert. But, how can such mental events as beliefs and desires cause John's mouth to move in such a way that he orders dessert?

## Sub-problems of Mental Causation

### Exclusion Problem

What follows is a summary of the causal exclusion problem in its simplest form, and it is merely one of several possible formulations.

To the extent that we do *not* have to go outside human physiology in order to trace the causal antecedents of any bodily movement, intentional action can be fully causally explained by the existence of these physiological antecedents *alone*. No mention of mental states need enter into the explanation. This troubles philosophers because intuitively it seems that mental states are crucial in causing a person to act (for example, their beliefs and desires). But, given that physiological facts are sufficient to account for action, mental states appear to be superfluous; they are at risk of being causally and explanatorily irrelevant with respect to human action (Yoo 2006, p. §3b.iii).

Many philosophers consider this apparent irrelevance to be a highly counter-intuitive and undesirable position to take. It ultimately leads to epiphenomenalism—the view that mental events or states are causally irrelevant, they are merely after effects that play no role in any causal chains whatsoever. Thomas Huxley famously noted that epiphenomenalism treats mental states like the steam coming off a train: it plays no causal role in the train's moving forward, it is merely an "emergent property" of the actual causation occurring in the engine (Walter 2003, p. §2).

## Problem of Anomalism

Another problem with mental causation is that mental events seem anomalous in the sense that there are no scientific laws that mental states can figure into without having exceptions. There are no "strict" laws, and mental events must factor into strict laws in order to fit respectably into the causal order described by current science.

In short, one response has been to deny that psychological laws involving mental states require strict, exceptionless laws. Jerry Fodor argues that non-basic (or "special") sciences do not in fact require strict laws (Fodor 1980). In current practice, special sciences (for example, biology and chemistry) have *ceteris paribus* laws (or laws with "all else being equal" clauses), according to which there are exceptions. However, only in the basic sciences (physics) are there strict, exceptionless laws. Thus, although mental states are anomalous, they can still figure into scientifically respectable laws of psychology.

## Problem of Externalism

In the latter half of the twentieth century externalism about meaning became espoused by many philosophers. Externalism is roughly the view that certain parts of an individual's environment play a crucial role in the meaning of at least some of an individual's words]. A thesis about meaning affects the mind insofar as our thoughts are *about* things in the world. A common view in the philosophy of mind is that at least certain mental states have intentional content in this sense. For example, one's belief that water is wet has the semantic content of *water is wet*. The thought is about water and the fact that it is wet. But, if externalism is true—if some of the contents of one's thoughts are constituted at least in part by factors external to one's mind—then there is yet another difficulty in explaining how mental states can cause physical states (Yoo 2006, p. §3b.ii)].

## Solutions to the Problem

There are several options for answers to the kinds of questions raised by the basic problem of mental causation.

## Dualist Solutions

Some have claimed that while the mental and the physical are quite different things, they can nonetheless causally interact with one another, a view going back to Descartes [(Descartes 1642/1986), especially meditations II & VI]. This view is known as *interactionist dualism*. The major problem that interactionist dualism faces is that of explicating a satisfactory notion of causation according to which non-spatial events, such as mental events, can causally interact with physical events. According to the current mainstream scientific world-view, the physical realm is *causally closed*, in that causal relationships only hold among physical events in the physical realm. Given these types of considerations, some argue that it is appropriate to say that the main assumptions in interactionist dualism generate the problem of mental causation rather than solve it.

## Physicalist Solutions

The other major option is to assert that mental events are either (at least contingently) identical to physical events, or supervene on physical events. Views that fall under this general heading are called physicalism or materialism. But, such views require a particular theory to explain how mental events are physical in nature. One such theory is behaviorism. Behaviorists, in general, argue that mental events are merely dispositions to behave in certain ways. Another theory is the identity theory, according to which mental events are (either type- or token-) identical to physical events. A more recent view, known as functionalism, claims that mental events are individuated (or constituted by) the causal role they play. As such, mental events would fit directly into the causal realm, as they are simply certain causal (or functional) roles. Finally, there is eliminative materialism, which simply denies that there are any such mental events; thus, there is really no problem of mental causation at all.

## Other Constructions

## Popper's Three-world Formulation

Related to dualism above, a more general and somewhat differently posed approach to mental causation is provided by Karl Popper's three worlds. Popper split the world into three categories:

1. The world of physical objects and events, including biological entities

2. The mental or psychological world, the world of our feelings of pain and of pleasure, of our thoughts, of our decisions, of our perceptions and our observations; in other words, the world of mental or psychological states or processes, or of subjective experiences.

3. The world of products of the human mind, including art, science, and religion.

World 3 includes physical theory as a particular case. But World 3 is a creation of the human imagination, and such acts of imagination are a part of World 2. Accordingly, one could argue that the physical notion of causality is a child of the imagination, and although causation has its successes in describing World 1, it may not apply to World 2 or World 3. The subjective aspects of theories contained in World 3 are not readily framed within the third-person perspective of science used to explain World 1.

From this perspective, it is hubris to suppose that the methods successful in describing World 1, in particular to suppose the notions of *cause* and *effect*, invented by World 2 in its creation of the theory of World 3 used to explain World 1, have direct application to Worlds 2 and 3 themselves, and control mental agency.

## Psychological Nativism

A still different approach to mental causation is based upon the philosophies of Kant, Chomsky and Pinker. These philosophers stress the impact of built-in aspects of mind, studied in the field of psychological nativism.

Immanuel Kant (1724–1804) pointed out that we all shape our experience of things through the filter of our mind, a view sometimes called epistemological solipsism. The mind shapes that experience, and among other things, Kant believed the concepts of *space* and *time* were programmed into the human brain, as was the notion of *cause* and *effect*. We never have *direct* experience of things, the *noumenal* world, and what we do experience is the *phenomenal* world as conveyed by our senses, this conveyance processed by the machinery of the mind and nervous system. Kant focused upon this processing. Kant believed in *a priori* knowledge arrived at independent of experience, so-called *synthetic a priori* knowledge. In particular, he thought that by introspection some aspects of the filtering mechanisms of the mind/brain/nervous system could be discovered. The following observations summarize Kant's views upon the subject-object problem, called Kant's Copernican revolution:

"It has hitherto been assumed that our cognition must conform to the objects; but all attempts to ascertain anything about these objects *a priori*, by means of conceptions, and thus to extend the range of our knowledge, have been rendered abortive by this assumption. Let us then make the experiment whether we may not be more successful in metaphysics, if we assume that the objects must conform to our cognition. This appears, at all events, to accord better with the possibility of our gaining the end we have in view, that is to say, of arriving at the cognition of objects *a priori*, of determining something with respect to these objects, before they are given to us. We here propose to do just what Copernicus did in attempting to explain the celestial movements. When he found that he could make no progress by assuming that all the heavenly bodies revolved round the spectator, he reversed the process, and tried the experiment of assuming that the spectator revolved, while the stars remained at rest. We may make the same experiment with regard to the intuition of objects."

*—Immanuel Kant, English translation by J. M. D. Meiklejohn of The Critique of Pure Reason (1781)*

Although Kant has posed the issue of built-in aspects of mind, the particulars that depend upon the science of his day have become outmoded. A more recent approach to these limitations is proposed by Noam Chomsky and Steven Pinker. Like Kant, Noam Chomsky raised the issue of the mind's inherent programming. Chomsky selected as a particular example the acquiring of language by children. Of course, language is indispensable in the formulation and communication of our perceptions of the objective world:

"People do not think in English or Chinese or Apache; they think in a language of thought. This language of thought probably looks a bit like all these languages;...But compared with any given language, mentalese must be richer in some ways and simpler in others."

*—Steven Pinker, The Language Instinct, p. 72*

Chomsky marshaled evidence that a child's rapid mastery of the complexity of language indicated an innate ability programmed into the development of the human mind from birth that could not be explained by the "blank slate" view of the infant mind. Rather, the mind has a built-in propensity to process symbolic representations. The origins of this ability were sought by Steven Pinker in a Darwinian struggle that established the survival value of the ability to communicate. According to Pinker, Charles Darwin himself "concluded that language ability is 'an instinctive tendency to acquire an art', a design that is not peculiar to humans but seen in other species such as song-learning birds." This observation is strongly supported by research on crows.

This work can be taken to suggest that although a physical theory is an intermediary between our observations and our notions of connections between them, it is an elaborate mental construction that is a meld of the way the mind works and objective observations. Although a physical theory is used to determine connections about objective events, the specific form of the theoretical construct is a product of subjective activities, and this particular form may well involve the workings of the brain. Perhaps some aspects of the universe's operation can be expressed in terms of mental constructs, but this process is analogous with the expression of a computer algorithm in terms of assembly language instructions peculiar to a particular computer, a translation by a compiler of the general statement of an algorithm into specific tiny steps that particular computer can handle.

From this standpoint, as with the philosophy of Kant, the first-person active actions of mental causation may involve innate workings of the brain itself.

## Supervenience

In philosophy, supervenience is an ontological relation that is used to describe cases where (roughly speaking) the upper-level properties of a system are determined by its lower level properties. Some philosophers hold that the world is structured into a kind of hierarchy of properties, where the higher level properties supervene on the lower level properties. According to this type of view, social properties supervene on psychological properties, psychological properties supervene on biological properties, biological properties supervene on chemical properties, etc. That is, the chemical properties of the world determine a distribution of biological properties, those biological properties determine a distribution of psychological properties, and so forth. So, for example, *mind-body supervenience* holds that "every mental phenomenon must be grounded in, or anchored to, some underlying physical base (presumably a neural state). This means that mental states can occur only in systems that can have physical properties; namely physical systems."

It is useful to know both when supervenience does and does not apply. For example, the value of a piece of paper money does not supervene on the micro-features of the paper it's made out of, because the value of the money is not just determined by internal features of the paper, but also by a broader distribution of social facts and institutions. The paper alone does not determine the value of the money, so supervenience does not obtain between the value and the paper (though it does obtain between the value of the money and some larger part of the world which includes the relevant social institutions). Some philosophical problems can be stated in terms of questions about supervenience. For example: do mental properties supervene on neural properties? (externalists in the philosophy of mind say "no", internalists say "yes.") Do the aesthetic properties of an art object supervene on physical properties of that object?

Though it is intuitively clear what supervenience is, formally specifying what it means involves a complex technical apparatus and a family of related but subtly different definitions. Most definitions involve comparisons of objects for indiscernibility. According to one standard definition, a set of properties A (e.g. mental properties) supervenes on a set of properties B (e.g. neural properties), if and only if any two objects x and y which share all properties in B (are "B-indiscernible") must also share all properties in A (are "A-indiscernible"). The intuitive idea is that if you could make a physical copy of a person, you'd also be making a psychological copy of that person. The reverse does not hold: two people could be in the same mental state, but that mental state could be supported by different brain states (the same mental state could be "multiply realizable" by different brain states). The properties in B are called the *base properties* (or sometimes *subjacent* or *subvenient properties*), and the properties in A are called the *supervenient properties*.

Supervenience has traditionally been used to describe relationships between sets of properties in a manner which does not imply a strong reductive relationship. For example, many hold that economic properties supervene on physical properties, in that if two worlds were exactly the same physically, they would also be the same economically. However, this does not entail that economics can be reduced in any straightforward way to physics. Thus, supervenience allows one to hold that "high-level phenomena" (like those of economics, psychology, or aesthetics) depend, ultimately, on physical substance, without assuming that one can study those high-level phenomena using means appropriate to physics.

## History

Supervenience, which means literally "coming or occurring as something novel, additional, or unexpected", from "super," meaning on, above or, additional, and "venire," meaning to come, shows occurrences in the Oxford English Dictionary dating back to 1844.

Its systematic use in philosophy is considered to have begun in early 20th-century meta-ethics and emergentism. As G.E. Moore wrote in 1922, "If a given thing possesses any kind of intrinsic value in a certain degree, then... anything *exactly like it*, must, under all circumstances, possess it in exactly the same degree". The usage also carried over into the work R. M. Hare. For discussion of the emergentist roots of supervenience see Stanford Encyclopedia of Philosophy: Supervenience.

In the 1970s Donald Davidson was the first to use the term to describe a broadly physicalist (and non-reductive) approach to the philosophy of mind. As he said in 1970, "supervenience might be taken to mean that there cannot be two events alike in all physical respects but differing in some mental respects, or that an object cannot alter in some mental respects without altering in some physical respects."

In subsequent years Terry Horgan, David Lewis, and especially Jaegwon Kim formalized the concept and began applying it to many issues in the philosophy of mind. This raised numerous questions about how various formulations relate to one another, how adequate the formulation is to various philosophical tasks (in particular, the task of formulating physicalism), and whether it avoids or entails reductionism.

## Definitions

In the contemporary literature, there are two primary (and non-equivalent) formulations of supervenience (for both definitions let A and B be sets of properties).

(1) A-properties supervene on B-properties if and only if all things that are B-indiscernible are A-indiscernible. Formally:

- $\forall x \forall y (\forall X_{\in B}(Xx \leftrightarrow Xy) \rightarrow \forall Y_{\in A}(Yx \leftrightarrow Yy))$

(2) A-properties supervene on B-properties if and only if anything that has an A-property has some B-property such that anything that has that B-property also has that A-property. Formally:

- $\forall x \forall X_{\in A}(Xx \rightarrow \exists Y_{\in B}(Yx \wedge \forall y(Yy \rightarrow Xy)))$

For example, if one lets A be a set of mental properties, B be a set of physical properties, and chooses a domain of discourse consisting of persons, then (1) says that any two persons who are physically indiscernible are mentally indiscernible, and (2) says that any person who has a mental property has some physical property such that any person with that physical property has that mental property.

Some points of clarification: first, the definitions above involve quantification over properties and hence higher order logic. Second, in (1), expressions of the form $(\forall X(Xx \leftrightarrow Xy))$ capture the concept of sharing all properties, or being indiscernible with respect to a set of properties. Thus, (1) can be understood more intuitively as the claim that all objects that are indiscernible with respect to a base set of properties are indiscernible with respect to a supervenient set of properties, or, as it is also sometimes said, that B-twins are A-twins. Finally, supervenience claims typically involve some modal force, however, the way that modal force is specified depends on which more specific variety of supervenience one decides upon.

(1) and (2) are sometimes called "schemata", because they do not correspond to actual supervenience relations until the sets of properties A and B, the domain of entities to which those properties apply, and a modal force have been specified. For modal forms of supervenience, the modal strength of the relation is usually taken to be a parameter (that is, the possible worlds appealed to may be physically possible, logically possible, etc.). Also, note that in the early literature properties were not always central, and there remain some who prefer to frame the relation in terms of predicates, facts, or entities instead, for example.

## Varieties of Supervenience

Beginning in the 1980s, inspired largely by Jaegwon Kim's work, philosophers proposed many varieties of supervenience, which David Lewis called the "unlovely proliferation". These varieties are based both on (1) and (2) above, but because (1) is more common we shall focus on varieties of supervenience based on it.

We can begin by distinguishing between local and global supervenience:

- Local: For any two objects x and y, if x and y are base-indiscernible, they are supervenient-indiscernible.

For example, if mental states locally supervene on brain states, then being in the same brain state entails being in the same mental state.

- Global: For any two worlds w1 and w2, if w1 and w2 are base-indiscernible, they are supervenient-indiscernible.

For example, if psychological properties globally supervene on physical properties, then any two worlds physically the same will be psychologically the same. The value of global supervenience is that it allows for supervenient properties to be determined not by local properties of an individual thing alone, but by some wider spatiotemporal distribution of things and properties. For example, something's being a dollar bill depends not only on the paper and the inks it is made out of, but also on a widely dispersed variety of features of the world it occupies.

Both local and global supervenience come in many forms. Local supervenience comes in strong and weak varieties:

- Weak: For any world w, and for any two objects x in w and y in w, if x and y are base-indiscernible, they are supervenient-indiscernible.
- Strong: For any worlds w1 and w2, and for any two objects x in w1 and y in w2, if x and y are base-indiscernible, they are supervenient-indiscernible.

The difference is essentially whether correlations between base and supervenient properties hold within actual worlds only, or across possible worlds. For example, if psychological properties strongly locally supervene on physical properties, then any two people physically the same, in any two worlds, will also be psychologically the same. On the other hand, if psychological properties only weakly locally supervene on physical properties, then those correlations between base and supervenient properties that hold in virtue of the supervenience relation are maintained within each world, but can be different in different worlds. For example, my physical duplicates in the actual world will have the same thoughts as I have; but my physical duplicates in other possible worlds may have different thoughts than I have in the actual world.

There are also several kinds of global supervenience relations, which were introduced to handle cases in which worlds are the same at the base level and also at the supervenient level, but where the ways the properties are connected and distributed in the worlds differ. For example, it is consistent with global mental–physical supervenience on the simple formulation described above for two worlds to have the same number of people in the same physical states, but for the mental states to be distributed over those people in different ways (e.g. I have my father's thoughts in the other world, and he has my thoughts). To handle this, property-preserving isomorphisms (1-1 and onto functions between the objects of two worlds, whereby an object in one world has a property if and only if the object which that function takes you to in the other world does) are used, and once this is done, several varieties of global supervenience can be defined.

Other varieties of supervenience include multiple-domains supervenience and similarity-based supervenience.

## Examples of Supervenient Properties

### Value Properties

The value of a physical object to an agent is sometimes held to be supervenient upon the physical properties of the object. In aesthetics, the beauty of La Grande Jatte might supervene on the physical composition of the painting (the specific molecules that make up the painting), the artistic composition of the painting (in this case, dots), the figures and forms of the painted image, or the painted canvas as a whole. In ethics, the goodness of an act of charity might supervene on the physical properties of the agent, the mental state of the agent (his or her intention), or the external state of affairs itself. Similarly, the overall suffering caused by an earthquake might supervene on the spatio-temporal entities that constituted it, the deaths it caused, or the natural disaster itself. The claim that moral properties can supervene upon non-moral properties is, however, disputed by proponents of a more restricted use of the concept.

## Mental Properties

In philosophy of mind, many philosophers make the general claim that the mental supervenes on the physical. In its most recent form this position derives from the work of Donald Davidson, although in more rudimentary forms it had been advanced earlier by others. The claim can be taken in several senses, perhaps most simply in the sense that the mental *properties* of a person are supervenient on their physical *properties*. Then:

- If two persons are indistinguishable in all of their physical properties, they must also be indistinguishable in all of their mental properties.

This is in direct contradiction with the probabilistic nature of the universe.

An alternative claim, advanced especially by John Haugeland, is a kind of *weak local* supervenience claim; or, weaker still, mere *global* supervenience. The claim that mental properties supervene globally on physical properties requires only a quite modest commitment: any difference between two possible worlds with respect to their instantiated mental properties entails at least *some* difference in the physical properties instantiated in those two worlds. Importantly, it does *not* require that the mental properties of an individual person supervene *only* on that person's physical state.

This weak global thesis is particularly important in the light of direct reference theories, and semantic externalism with regard to the content both of words and (more relevant to our concerns here) of *thoughts*. Imagine two persons who are indistinguishable in their local physical properties. One has a dog in front of his eyes and the other has a dog-image artificially projected onto his retinae. It might be reasonable to say that the former is in the *mental state* of seeing a dog (and of knowing that he does so), whereas the latter is not in such a state of seeing a dog (but falsely believes that he sees one).

There is also discussion among philosophers about mental supervenience and our experience of duration. If all mental properties supervene only upon some physical properties at durationless moments, then it may be difficult to explain our experience of duration. The philosophical belief that mental and physical events exist as a series of durationless moments that lie between the physical past and the physical future is known as presentism, and is a form of belief in Galilean relativity.

## Computational Properties

There are several examples of supervenience to be found in computer networking. For example, in a dial-up internet connection, the audio signal on a phone line transports IP packets between the user's computer and the Internet service provider's computer. In this case, the arrangement of bytes in that packet supervenes on the physical properties of the phone signal. More generally, each layer of the OSI Model of computer networking supervenes on the layers below it.

We can find supervenience wherever a message is conveyed by a representational medium. When we see a letter "a" in a page of print, for example, the meaning *Latin lowercase "a"* supervenes on the geometry of the boundary of the printed glyph, which in turn supervenes on the ink deposition on the paper.

## Biological Properties

In biological systems phenotype can be said to supervene on genotype. This is because any genotype encodes a finite set of unique phenotypes, but any given phenotype is not produced by a finite set of genotypes. Innumerable examples of convergent evolution can be used to support this claim. Throughout nature, convergent evolution produces incredibly similar phenotypes from a diverse set of taxa with fundamentally different genotypes underpinning the phenotypes. One example is evolution on islands which is a remarkably predictable example of convergent evolution where the same phenotypes consistently evolve for the same reasons. Organisms released from predation tend to become larger, while organisms limited by food tend to become smaller. Yet there are almost infinite numbers of genetic changes that might lead to changes in body size. Another example of convergent evolution is the loss of sight that almost universally occurs in cave fish living in lightless pools. Eyes are expensive, and in lightless cave pools there is little reason for fish to have eyes. Yet, despite the remarkably consistent convergent evolution producing sightless cave fish, the genetics that produce the loss of type phenotype is different nearly every time. This is because phenotype supervenes on genotype.

## Arguments Against Supervenience-based Formulations of Physicalism

Although supervenience seems to be perfectly suited to explain the predictions of physicalism (i.e. the mental is dependent on the physical), there are four main problems with it. They are *Ephiphenomenal ectoplasm*, the *lone ammonium molecule problem*, *modal status problem* and the *problem of necessary beings.*

## Epiphenomenal Ectoplasm

*Epiphenomenal ectoplasm* was proposed by Horgan and Lewis in 1983, in which they stated, a possible world (a world that could possibly exist) *W* is identical to our world in the distribution of all mental and physical characteristics (i.e. they are identical), except world *W* contains an experience called *epiphenomenal ectoplasm* that does not causally interact with that world. If supervenience physicalism is true, then such a world could not exist because a physical duplicate of the actual world (the world that is known to exist) could not possess an *epiphenomenal ectoplasm*. This was rectified by Frank Jackson, by adjusting the application of supervenience within physicalism to state "Physicalism is true at a possible world *W* if and only if any world which is a minimal physical duplicate (i.e. identical) of *W* is a duplicate of *W simpliciter*."

### The Lone Ammonium Molecule Problem

The *lone ammonium molecule problem* provides a problem for Jackson's solution to *epiphenomenal ectoplasm*. It was proposed by Jaegwon Kim in 1993 when he stated that according to Jackson's idea of supervenience, a possible world *W* was identical to the actual world, except it possessed an extra ammonium molecule on one of Saturn's rings. This may not seem to provide much of a problem, but because Jackson's solution refers only to minimal physical duplicates, this allows for the mental properties of *W* to be vastly different from in the actual world. If such a difference would cause mental differences on Earth, it would not be consistent with our understanding of physicalism.

### Modal Status Problem

The *modal status problem* is only problematic if one thinks of physicalism as a contingent truth (i.e. not necessary truth), because it is described in terms of modal notions (i.e. through modal realism). The problem is presented when from the statement "Minimal physical truths entail all truths", one derives the statement "*S* (a statement that describes all minimal physical truths) entails *S** (a statement that describes the world)". This statement is a necessary truth, and therefore supervenience physicalism could not be contingent. The solution to this is to accept the above statement not as the equivalent of physicalism, but as an entailment of it.

### Problem of Necessary Beings

The Problem Of Necessary Beings Was proposed by Jackson in 1998, in which he stated that a necessary being exists in all possible worlds as a non-physical entity, and therefore proves physicalism false. However, physicalism allows for the existence of necessary beings, because any minimal physical duplicate would have the same mental properties as the actual world. This however is paradoxical, based on the fact that physicalism both permits and prevents the existence of such beings. This violates Hume's fork which states, "there are no necessary connections between distinct existences".

## Cartesian Theory

Descartes is one of the classical founders of non-computational theories of mind. Without a proper understanding of Descartes's view on mind, it is impossible to discuss contemporary philosophy of mind. This paper deals with two important issues, namely, the existence of mind and its nature, and how Descartes's idea of mind is non-computational. I shall argue that Williams, Hintikka, Malclom, and many others philosophers philosophical arguments will not cope with Descartes notion of mind because the way they are defining the notion of mind is negating the existence of mind and also its nature. Secondly, I argue that Descartes idea of mind is non-computational because the way Ryle, Quine, and other functionalists or founder of cognitive scientists defined it is completely mechanical or behavioural and to which the notion of computationally is applicable, and the mental qualities are credibility to machines. My main aim in this paper is to clarify Descartes's notion of mind from subjective point of view. I believe that Descartes's notion of mind cannot be explained or characterized in a computationalistic approach and that are the subjective mental states which we can see from the first-person perspective of their proper understanding.

## Cartesian Mind and Its Nature

According to Descartes, to know something implies that there is a mind, i.e., the existence of a knowing subject means that there is a mind. Again, he tries to find out through his *cogito* argument that there is at least one knowing subject, i.e., his own self. He arrives at this truth through his method of doubt. The method led Descartes to argue that the whole body of knowledge might be mistaken. In Descartes's words, "I will suppose then, that everything I see is spurious. I will believe that my memory tells me lies, and that none of the things that it reports ever happened. I have no sense. Body, shape, extension, movement and place are chimeras"

In this context, Descartes raises a fundamental question: from the possible non-existence of the external world and our own bodies, does it not follow that it is possible that we ourselves do not exist? Again, his reply to the above question is as follows, "No: if I conceived myself of something then I certainly existed. But there is a deceiver of supreme power and cunning who is deliberately and constantly deceiving me. In that case, I too undoubtedly exist, if he is deceiving and let him deceive me as much as he can, he will never bring it about that I am nothing so long as I think that I am something. So after considering everything very thoroughly, I must finally conclude that this proposition, *I am, I exist*, is necessarily true whenever it is put forward by me or conceived in my mind."

Thus to think that one does not exist, one must exist. Therefore one's own non-existence is inconceivable. For Descartes, '*cogito ergo sum*' is an indubitable proposition. Doubting one's own existence presupposes one's existence. Now the question arises: What is the nature of the statement '*cogito ergo sum*'? Is it a syllogistic inference like, 'whatever thinks exists; I think; therefore, I exist'? For Descartes, it is not a syllogistic inference; it is rather a self-evident truth known "by a simple intuition of the mind."[3] Thus scholars are divided among themselves as to the exact nature of the transition from '*cogito*' to '*sum*'. As Bernard Williams has shown, there is something unique about the '*cogito*' which cannot be replaced by any other verb, for instance, '*ambulo*'. '*Ambulo ergo sum*' is not as self evident as '*cogito ergo sum*'.

Moreover, unlike Williams, Hintikka argues that *cogito ergo sum* is not an inference but a performance. He says, "The function of the word 'cogito' in Descartes' dictum is to refer to the thought-act through which the existential self verifiability of 'I- exists' manifests itself." For him, the relation of '*cogito*' to '*sum*' is similar to the relation of a process to its product. The truth of 'I exist' is revealed to one only when one actively thinks just as there is illumination only when the source of light exists. The truth of 'I exist' cannot be revealed by any arbitrary human activity such as breathing, etc. but only by thinking. An attempt to think one's own non-existence amounts to persuading oneself to the belief that one does not exit. Through each thought, including the thought of one's own non-existence, the truth of '*sum*' is verified. The self's coming to know its own existence and it is revealed in the act of thought.

According to Descartes, the thought act is due to the thinking thing, which is the self. For him, again, the thinking thing or the self is that which, "But what then am I? A thing that thinks. What is that? A thing that doubts, understands, affirms, denies, is willing, is unwilling, and also imagines and has sensory perceptions". The existence of the thinking thing is the same as the existence of the knowing thing. From this statement it follows that there is a mind, which has the power of knowing something. And if there exists at least one mind, it is logically and even empirically possible that there are other minds.

Now the question is: If there is/are mind(s), what is its nature or essence? Thought, according to Descartes, is the essence of mind. The essence of a thing is defined as that which is necessary for its existence, Descartes claims that he has clear and distinct perception or awareness that he is a thinking thing and nothing other than thought belongs to his nature.

Malcolm argues that in identifying thought as mind's essence Descartes employs the following principle: "X is my essence if it is the case that (a) if I am aware of X, then (necessarily) I am aware of myself and (b) if I am aware of myself then (necessarily) I am aware to X. thinking satisfies these conditions. Ergo, thinking is my essence."

Malcolm illustrates how thought alone satisfies the single principle that any act of thought, for Descartes, is identical with the act of consciousness. Consequently, if I am aware of anything, then I am thinking, and so if I am aware of thinking then I am thinking; and if I am thinking I am aware of thinking. In Malcolm's view, though Descartes does not explicitly maintain that whatever I think, I am aware of myself, he would be drawn to accept it partly because " . . . the best support for his principle 'I think ergo I exist' is at the same time a support for the principle ' I think ergo I am,' 'I am aware that I exist'." So thought satisfies the condition (a) of the above principle. It also satisfies the condition (b), as every act of awareness of myself is also an act of my awareness of something other than myself. Since acts of thought are identified with acts of consciousness, it follows that cognitive acts are conscious acts. So far, for Descartes, one of the most important aspects of cognitive states and processes is their phenomenality. Our perception, understanding, judgment etc. can be defined and explained only in relation to consciousness. According to Descartes, the mind is a thinking substance endowed with various faculties, such as sensory perception, understanding, willing, etc. For him, it is one and the same mind, which wills, understands and has sensory perceptions.

Moreover, Descartes grants that mind is associated with body, and mind provides metaphysical support to the body. This drives him to the examination of the nature of the body in its metaphysical aspects i.e., body in the most general sense of the term. The most general conception of body attained through a clear and distinct perception of the intellect, is that it is an extended substance, a continuum with three dimensions of length, breath and height. As in the case of the mental substance, the extended substance too is known through its acts or modes which, according to Descartes, are shape, size, position, motion, rest, etc.

## Descartes's Idea of Mind is Non-Computational

Till now we have discussed the Cartesian mind and its nature. In the Cartesian scheme of mind, there is no place for computationality. Because the thought act is due to the subjective thinking thing, which is the self. Again, this subjective thinking thing or the self is that which " doubts, understands, affirms, denies, is willing, is unwilling, and also imagines and has sensory perceptions." The existence of the thinking thing is same as the existence of the subjective thinking thing, because it is the subject, who thinks. All these subjective activities are non-computational because the subjective activity is not mechanical. The mental processes, for Descartes, are intentional and are the free acts of the thinking subject. Hence they cannot be mapped mechanically in an algorithmic system.

Descartes concept of 'I think' presupposes subjective experience, because it is 'I' who experience the world. Descartes' notion of 'I' negates the notion of computationality in the mind. The essence of mind is thought, and the acts of thoughts are identified with acts of consciousness. Therefore, it follows that cognitive acts are conscious acts, but not computational acts. Thus for Descartes, one of the most important aspects of cognitive states and processes is their phenomenality because our judgments, understanding, etc. can be defined and explained only in relation to consciousness, not in relation to computationality. We can only find computationality in machines, and not in the mind, which wills, understands and judges. Descartes' dictum, "I think, therefore, I am" not only establishes the existence of the self which thinks and acts but also its freedom from mechanistic laws, to which the human body is subject.

Moreover, when Descartes makes the distinction between mind and body, he did not say that the idea of the mind is that of a ghost, although he did say that the idea of body is that of a machine. Ryle in his book, '*The Concept of Mind*' says that Descartes's distinction between mind and body is a myth. He argues, "I shall often speak of it, with deliberate abusiveness, as 'the dogma of the ghost in the machine'. I hope to prove that it is entirely false, and false not in detail but in principle." According to Ryle, Descartes's distinction between mind and body commits a category-mistake.

As Ryle said, "my destructive purpose is to show that a family of radical category mistake is the source of the double-life theory. The representation of a person as a ghost mysteriously ensconced in a machine derives from this argument. Because, as is true, a person's thinking, feeling and purposive doing cannot be described solely in the idioms of physics, chemistry and physiology, therefore they must be described in counterpart idioms. As the human body is a complex organized unit, so the human mind must be another complex organized unit, though one made of a different sort of stuff and with a different sort of structure. Or, again, as the human body like any other parcel of matter, is a field of causes and effects, so the mind must be another field of causes and effects, though not (Heaven be praised) mechanical causes and effects."

In Ryle's understanding of mind, mind becomes as much mechanical as the body and is therefore non-different from the body. However, Descartes refutes the mechanistic reading of mind. As we have seen, Descartes is a dualist, rather than a mentalist. Descartes's argument for the mind, which is distinct from body, needs to be understood as an argument for the logical possibility of their separate existence and not for the fact that they exist independent of each other. The separability argument is as follows: "First, I know that everything, which clearly and distinctly understands is capable of being created by God so as to correspond exactly with my understanding of it. Hence, the fact that I can clearly and distinctly understand one thing apart from another is enough to make, certain that two things are distinct, since they are capable of being separated at least by God. The question of what kind of power is required to bring about such a separation does not affect the judgment that the two things are distinct. Thus, simply by knowing that I exist and seeing at the same time that absolutely nothing else belongs to my nature or essence except that I am a thinking thing, I can infer correctly that my essence consists solely in the fact that I am a thinking thing. It is true that I may have (or, to anticipate, that I certainly have) a body that is very closely joined to me. But nevertheless, on the one hand, I have a clear and distinct idea of myself, in so far as I am simply a thinking, non-extended thing; and on the other hand I have a distinct idea of body, in so far as this is simply an extended non-thinking thing. Accordingly, it is certain that I am really distinct from my body, and can exist without it."

Descartes has already proved in the *Second Meditation* the existence of a thinking being who has a clear and distinct perception of mind as a thinking, non-extended thing. This is a proof of the non-mechanical mind which is different from the body subject to mechanical laws. Similarly, in the *Fifth Meditation*, he has shown, that he has a clear and distinct idea of a body as extended and a non-thinking substance. This is to suggest that the mechanically existing body is ontologically distinct from the non-computational mind.

The above distinction between mind and body supposes that there is no 'ghost' in human body or 'ghost in the machine'. Descartes did not admit the existence of ghost in the machine. Had Descartes admitted that there was a ghost in the human body, then the mind itself would become computational, and there would be no necessary distinction between mind and body? Because the ghost itself is a body. But Descartes admits the distinction between mind and body and this shows that the mind is non-computational. It is mind, which has the capacity of intelligence, and understanding. The Cartesian way of understanding of the concept of intelligence is anti-physicalist and anti-behaviourist and hence is anti-computational.

The human mind is beyond the sphere of computationality, because the human mind has innate ideas, which are embedded as the innate dispositions of the human mind. These ideas are *a priori* in the human mind and are the basic in-born propensities. Descartes observes, "My understanding of what a thing is, what truth is, what thought is, seems to derive simply from my own nature. But my hearing a noise, as I do now, or seeing the sun, or feeling the fire, comes from thing which are located outside me, or so I have hitherto judged. Lastly, sirens, hippogriffs and the like are my own invention."

The above observation of Descartes shows that innate ideas are not produced in us by the senses. If the ideas were conveyed to us by the senses like heat, sound, etc., we would not have to refer to anything outside ourselves, they too would be innate. For Descartes, "the ideas of pain, colours, sounds, and the like must be all the more innate if, on the occasion of certain corporeal motions, our mind is to be capable of representing them to itself, for there is no similarity between these ideas, and the corporeal motions." Here it follows that there is a distinction between innate and adventitious ideas and that innate ideas are universal ideas, whereas adventitious ideas are particular ideas. As Descartes points out that hearing a noise, seeing the scene, and feeling the fire are all particular ideas. Again it must be noted that the perception of the particular is not possible without the universal. Innate universal ideas are a necessary requirement for the cognition of the particular objects in the world.

Following Descartes, Chomsky established that language too is an innate faculty of the human species. Language becomes the essence that defines what it is to be human. Language is purely a syntactic system, according to Chomsky, and it therefore has a logical form which is universal and innate world. Language must also have an essence; something that makes language what it is and inheres in all languages. That essence is called 'universal grammar'. Language does not arise from anything bodily. Studying the brain and body can give us no additional insight into language. The basic tenets of Chomsky's linguistics are taken directly from Descartes. The only major tenet of Descartes that Chomsky rejects is the existence of the mental substance different from the human brain. Chomsky accepts that the human brain embodies the innate grammatical structures.

Like Chomsky, Quine also affirms that there can be no philosophical study of mind outside psychology: progress in philosophical understanding of the mind is inseparable from progress in psychology. Because, psychology is a 'natural science' studying a natural phenomena, that is, a physical human subject. Quine argued, "a dualism of mind and body is an idle redundancy." Quine holds that "corresponding to every mental state, however fleeting or however remotely intellectual, the dualist is bound to admit the existence of a bodily state that obtains when and only when mental one obtains. The bodily state is trivially specifiable in the dualist's own terms, simply as the state of accompanying a mind, that is in the mental state. Instead of ascribing the one state to the mind, then, we may equivalently ascribe the other to the body. The mind goes by the bound, and will not be missed."

Quine's position is that there are irreducible psychological properties, but all explanation is ultimately physical. His account of our mental concepts emerges as he examines how we acquire them, how we learn, etc. He explains, "such terms are applied in the light of publicly observable symptoms: bodily symptoms strictly of bodily states, and the mind is as may be. Someone observes my joyful or anxious expression, or perhaps observes my gratifying or threatening situation itself, or hears me tell about it. She then applies the word 'joy' or 'anxiety'. I find myself applying those words to some of my subsequent states in cases where no outward signs are to be observed beyond my report itself. Without the outward signs, to begin with, mentalistic terms could not be learned at all." Quine opposes the Cartesian dualism and therefore arrives at a behaviourist and functionalist conception of mind. He reduces the mental states like beliefs and other propositional attitudes to functional states.

If both Chomsky and Quine are right about the nature of mind, then Descartes's view of mind is wrong. That is, if that human brain is the cause of the mental states, then we cannot but arrive at the conclusion that the mental states are causally computable within a physical system. Chomsky and Quine define the mental qualities in terms of physical qualities. Therefore, they define mind in terms of the computational functions of the brain. However, Descartes claims that all ideas in the mind are mental representations.

In the *third meditation*, Descartes gives an extensive account of ideas. He says, "thus when I will, or am afraid, or affirm, or deny, there is always a particular thing which I take as the subject of my thought, but my thought includes something more than the likeness of that thing. Some thoughts in this category are called volitions or emotions, which others are called judgments." The above quotation shows that some thoughts are images of things, i.e., they represent things in the world. That is, they have an object or content by which they are individuated as an idea of this particular thing or being.

Descartes also considers an 'idea' to refer to the 'form' of any thought. In his words, "I understand this term to mean the form of any given thought, the immediate perception of which makes me aware of the thought. Hence, whenever I express something in words, and I understand what I am saying, this very fact makes it certain that there is within me an idea of what is signified by the words in question." The ideas, for Descartes' arte thus representational and intentional in character.

Descartes, unlike Hobbes and Gassendi, is not a naturalist and keeps the thought content free from naturalization to which Hobbes and Gassendi are committed. For them, thoughts are mechanical processes in the brain. In reply to Descartes, Gassendi says,

"I thus realize none of things that the imagination enables me to grasp is at all relevant to this knowledge of myself which I posses, and that the mind must therefore be most carefully diverted from such things if it is to perceive its own nature as distinctly as possible." On the contrary, Descartes holds that individual acts of imagination, is as much as they are experiences are relevant to grasping the nature of mind, because the mind is a thinking thing free from the mechanistic processes of the brain. What separates Descartes' dualism from contemporary functionalism and identity theories is not so much his distinction between an immaterial mind and extended material body as his notion of the human being as a unity of mind and body, with the properties not reducible to either mind or body, but dependent precisely on their 'substantial' union.

What Descartes is drawing attention here is firstly, no machine could have the capacity to use linguistic and other signs to express thoughts and to give appropriate responses to meaningful speech, and secondly, machine could not have the capacity to act intelligently in all sorts of situation. Here, animal communication have not offered counter evidence to Descartes's assumption because human language is based on an entirely distinct principle, nor has modern linguistic dealt with his observation in serious way. For the 'Cartesian' tradition in linguistic are the idea of an innate, universal grammar and the idea that the study of the structure of this argument will revel the structure of thought or mind.

Descartes's argument that brutes or machines cannot think in the light of the general question of what makes an utterance or a symbolic structure meaningful. The kind of automatic, rule governed computation or symbol processing that a Turing machine instantiates and that can be performed by electronic computers would not count as thinking in Descartes sense: nor would the mechanical operations of a computer or robot, no matter how ingenious or intelligent, count as rational behaviour as he understands it. Not only is much a view if thinking too narrow, it is based on precisely the kind of category mistake that Ryle attributes to the Cartesians which I have discussed but Descartes himself is not guilty of explaining thought in terms of non- extensions only.

As Pradhan clarifies that Descartes is not a reductionist as he feels that mind cannot be reduced to anything else and it must have an autonomous existence alongside the existence of the material body. The 'I think' of the mental reality does not deny its 'I exist' character in the world; rather it is an affirmation of it. In that sense, we cannot say that Descartes subjectivized the mental world and thus made it into a private world. He made every effort to keep an objective constraint on the subjective mind and thus forestalled all ascetical questions about the existence of other minds. This is because Cartesian doctrine of the mind and its inner experience does assume that we know other minds as much as we know our own. That is the reason why Descartes called the 'I think' the absolute basis of all our knowledge-claims about others and the external world. Thus the self or mind is irreducible/not explainable in terms of the body or machines whether mine or another's. In view of this we can say that the Cartesian philosophy of mind not based on a mistake and that it has shown the right way to the understanding of the mind.

Of course Descartes would not have accepted the idea of mechanical or computational, artificial intelligence, he may still be considered an important forerunner of cognitivist and computational view of mind. Because the essence of the mind is rational thinking and that rational thought or cognition can be studied independently of other phenomena, like sensation, and emotions that Descartes stated that body dependent on mental phenomena, to which mind refer as 'consciousness'. Although Descartes did not identify mental thought with consciousness, emotions, awareness, etc, but regarded that all those are condition of thought. While arguing the existence of mind, Descartes talk about that the mind acting in some particular location in the brain to contemporary literal talk about mental processes as computational activity in the brain. But Descartes would not have accepted the mechanical application of rules on syntactic structures as a sufficient condition for rational symbol manipulation. The kind of automatic, rule-governed computation or symbol processing that a Turing machine instantiates and that can be performed by electronic computers would not count as thinking from the Cartesian point of view. Because Cartesian thinking is neither reducible to a narrowly understood rational capacity nor to consciousness. He clearly mentioned that consciousness is a necessary condition for thought.

5

# Subjective Analysis of Mind

The mind cannot be explained in only neurological or objective terms. The following section studies the subjective nature of the mind. It focuses on questions which are very fundamental in nature. This section helps the readers in developing a better understanding of the subjective aspect of mind.

## Philosophical Concepts

The concept of person is one of the most important concepts in the philosophy of mind. The present paper's aim is to outline and explain the non-materialist theory of the mind and person. Here the fundamental question is, what is a person? And what is its nature? It was Descartes who has proposed a theory of mind and person according to which a person is not just a material body, but also a thinking self. According to Descartes, a person is a self, a self-conscious mind which thinks, feels, desires and so on. The materialists have, however, rejected the Cartesian theory of persons, and have argued that persons are just material bodies, though they are complex material systems with some sort of mental properties.

However, it is wrong to say that mind is the brain or mind has only physical properties. Our brain has a particular size, shape and spatial location. In virtue of these qualities, our brain has a particular look. We can say that our brain can be variously experienced. The qualities of such experiences are related in some way to the material object. But, if this is so, where do we situate the qualities of experience? A neuroscientist says that all these are neural activities. Now the question is; where are they? The answer is that they are located in our mind. This implies that the mind is distinct from the body. The problem of this essay is the question, are persons material bodies? The materialists have argued that persons are material bodies, although very complex material bodies. My aim is to refute person as a material entity and establish that person is a non-material entity.

### What is a Person?

In this paper my intention is to argue for a non-materialistic view of person. That is to say, this paper intends to examine the nature of person from a non-materialist point of view. Before analyzing this concept of person, we have to raise a few questions like what is a person, what is the nature of person, and so on. These questions are fundamental questions in the philosophy of mind. In fact the English word 'person' is alleged to have been derived from the Latin 'persona', which was the mask worn by actors in dramatic performances. Neither in common usage nor in philosophy has there been a univocal concept of 'person'. In common usage 'person' refers to any human being in a general way. The person is distinct from a thing or material object. It generally stands for a living conscious human being.

Moreover, Strawson's definition of person is different from Williams. Strawson defines 'person' as "a type of entity such that both predicates ascribing states of consciousness and predicates ascribing corporeal characteristics, a physical situation, etc., are equally applicable to a single individual of that single type." Thus, for Strawson, persons are unique individuals who have both mental and physical attributes. Thus persons are neither purely physical body, nor are they pure spiritual substances. However, from Strawsonian view of persons is purely non-material, where as Williams view of person is purely material, which opposes Strawsonian's view. This is because Williams claim is that bodily continuity is a necessary condition for personal identity, because according to Williams, it is body which identify the persons, but not the mind, and there is no mind at all, therefore, bodily criterion identify the persons.

Thus, it is clear that Strawson would certainly reject the contention that mental attributes are reducible to physical attributes because Strawson admit that the concept of persons non-material. But here the question arises, does Strawson wish to say that person are bodies of a certain sort, namely, bodies which have mental attributes as only?

Strawson holds that persons have bodily attributes too. But unlike ordinary bodies, persons are things, which have mental attributes as well. According to Strawson, it is essential to persons that they be entities, which necessarily have both mental and bodily attributes. In addition, those mental things are essentially different from physical things? They are different types of substance. Persons are radically different material bodies. Strawson's theory looks like dualistic in holding that there are two different types of subjects, the physical bodies and persons.

Again, physical bodies necessarily have only one dimension, that is, physical dimension. Persons necessarily have two dimensions, a physical and a mental dimension. Persons thus have a dual nature. Now, we have to look at the relation between the knowledge a person has of himself and the knowledge that others have of him. Moreover, if the unity of a person is necessarily connected with the continuance of his body through time, then it is impossible for a person to survive the death of his body. Secondly, if bodily identity is a necessary criterion of personal identity, then it could not be shown that some non-physical characteristics of a person continues after his bodily death. On the other hand, if bodily identity is not a necessary criterion of personal identity, perhaps bodily death is merely one major event in a person's history and not the end of his life. Finally, if the fundamental criterion of identity were memory, it would follow that a person might be known to have survived death because he continued to have memories in his disembodied state.

The most important fact about the person is the self. The self is sometimes used to mean the whole series of a person's inner mental states and sometimes the spiritual substance to which they belong. The self does not refer to the body but to the mental history of the person. This made the unity problem seem intractable, when the mental images, feelings, and the like, are contrasted with the temporal persistence. In Strawson's sense, a person is a thing which necessarily has both mental and physical aspects. The person is primarily the subject of mental experience. In the person theory we cannot say that a person is a body, but we can say that a person is, in part, a body. If person is a body, then it cannot be a conscious mind. One of the important questions is can we even say that a person has a body? Shaffer supposes that Strawson would want to be able to say that. But what would it mean about the theory of person? It means that persons have bodily attributes. Another question is, does it say anything about the relation between a person and a body? The body necessarily has bodily attributes and has nothing to do with a person's attributes. But Strawson's view is that persons have both bodily and mental attributes.

We recognise all human beings as persons. This is because we generally do not make a distinction between persons and human beings. But we can hardly contemplate the existence of biologically very different persons inhabiting other planets, who are not human beings like us. However, the concept of person is in someway an ineliminable part of our conceptual scheme. In our conceptual scheme person and human beings coincide.

Joseph Margolis in his book on '*Persons and Minds*' mentioned that persons are the particulars that have minds and nervous systems, sensation and brain processes. But this will not quite do. A nervous system is not a person, nor is a psyche a person. It is at once the subject of both neurological and psychological predicates. In other words, it is both a nervous system and a psychic entity. Persons are not meriologically complex entities nor kind, each of which contains parts, a non-physical basic subject and a purely corporeal object to which this subject is in some way attached; for such a claim would not allow us to ascribe psychological attributes or corporeal attributes to the person as a whole. It is because persons are more than their bodies and that they are not reducible to any kind of body gross or subtle. The person-substance as described above is not taken to exclude the material properties as such. They only exclude the fact that persons are material bodies and nothing else. Persons are autonomous so far as their description in terms of bodies and mind is concerned. But it is not that no reference to body and mind is to be retained at all. Thus, persons description have the attributic reference to body and mind.

From the above discussion we can reiterate the Cartesian distinction between the mind and the body. They are opposed to each other because the essence of mind is thinking and the essence of body is extension. That is to say, the body is something spatial which is perishable. Moreover, the mind or person is something non-spatial. After death, the body only remains. This concept of the body becomes gruesomely explicit when we refer to it as 'the remains'. However, it is this conception of the body, which comes closest to that found in the person theory. In this theory we find that the body is not a person, nor is it a part of a person. It is the person, insofar as he is thought of as the subject of bodily attributes. But it becomes a reality at death. We call it a corpse.

Therefore, one of the paradoxical implications of the person theory is that the body which a person has, cannot be conceived of as a physical object subject to the law of physical world. As we know from this theory, that persons are conscious. Finally, from the above examination we came to know that a person's body is not a physical thing.

## Person, Mind and Consciousness

As we have argued so far, person is an entity which has both mental and physical attributes. Hence, we could say of a person that he is five feet tall, weight one hundred kilograms, etc. But more importantly we could say, that he is thinking about his friends, feels a pang of happiness or is sad, or so on. We may therefore say that person has a mind, which is different from his body, because the subject of consciousness does not mean a body of a certain sort. But it still might turn out that whatever is a subject of consciousness is identical with a body of certain sort.

However, Strawson rejects the view that the subject of a state of consciousness is wholly immaterial, non-physical, a thing to which nothing but states of consciousness can be ascribed. According to him, consciousness is not applicable to physical things, nor to purely immaterial substance is applicable to person. But a fundamental question is, what is consciousness? Generally, consciousness is described as something, which distinguishes man from a good deal of the world around. Only a person possesses this consciousness, which is not by other material objects. Again, the question arises, what is this consciousness which a person certainly has, but rocks and other animate beings do not have? As G.E. Moore writes, "The moment we try to fix our attention upon consciousness and to see what distinctly it is, it seems to vanish: it seems as if we had before us as mere emptiness when we try to introspect the sensation of blue, all we can see is the blue; the other element is as if it were diaphanous."[3] Of course, we know perfectly well that we are conscious of things around us including other people, but we do not grasp consciousness itself.

However, it is this common feature, consciousness, which may be said to be the central element in the concept of mind. Shaffer points out that, if we were asked to give a general characterization of the branch of philosophy called philosophy of mind, we might say that it is that branch particularly concerned with the nature of consciousness, we will call them "mental phenomena" to which only beings capable of consciousness are subject, mental phenomena are such as ways of being conscious, that is, hearing, imaging, etc. Person is being minded have the capacity of doing the mental activities. Such activities include thinking, willing, feeling, understanding, speaking, communicating, and above all, remembering the past. Mental activities are such that they presupposes that there is a thinker who is capable of these activities. The thinker is here a subject or 'I' who is or has the capacity of consciousness. Wherever we will find the concept of 'I', we will find the existence of consciousness because it is person who stands for the concept of 'I', have consciousness.

One of the most general views is that the philosophy of mind is concerned with all mental phenomena which they themselves are concerned with consciousness. Philosophers from Descartes onwards have accepted consciousness as a fundamental metaphysical reality. I remain the same person if I am conscious of being so, even though my body should change drastically and be diminished through amputation. Logically, it is possible that I should remain the same person although I am altogether disembodied. Persons are indivisible, non-corporeal simple entities. It is because it becomes difficult here to distinguish persons so construed from metaphysical selves, transcendental egos, spirits, mental substances, souls, and other similar immaterial substances. However, the concept of person does not fit into these entities, because persons are, if anything, concrete beings in the world. One can ascribe consciousness to others only if one can identify other subjects of experience. In addition, one cannot identify other subjects if one can identify them only as subjects of experience, possessors of state of consciousness. The latter must have concrete existence in the world.

If we are too obsessed with the 'inner' criteria we shall be tempted to treat persons essentially as minds. However, admitting outer criteria does not mean that there are no states of consciousness. We should claim that some P-predicates refer to the occurrence of state of consciousness. The persons are uncertainly identifiable beings having a life of their own. They are not definitely Cartesian egos; rather they possess a mixed bag of M-predicates and P-predicates. Persons are in any case conscious individuals who can be ascribed a large number of P-predicates such as thinking, feeling, willing, deciding, etc. these conscious states, according to Searle, are intentional, that is, are of something. That is, they are directed at something outside them. Thus, persons who have these conscious states are intentional and mental beings.

Again, only a being that could have conscious intentional states could have intentionality at all, and so every unconscious intentional state is at least potentially conscious. This thesis has enormous consequence for the study of the mind. But there is a conceptual connection between consciousness and intentionality that has the consequence that a complete theory of intentionality requires an account of consciousness. And our consciousness is consciousness of something. Thus persons have the essential feature of consciousness. There is an interconnection between person, mind, and consciousness. Empirically, there is distinction among them. But transcendentally, they point in the same direction. It is right to say that a person is a mental being, and the essence of mind is consciousness. Therefore, the concept of the mind, the person, and the consciousness go together. Thus consciousness is related to mind, which also belongs to a person.

## The Dual Nature of Person

The problem of person has traditionally been raised in a dualist context. It has been greatly influenced those who have discussed by the picture of a person as composed of two entities, body and mind, which are contingently related to each other. However, the person-substances are not merely a set of properties, physical or mental, because they are not fully exhausted in their descriptions. The descriptions of the persons as having such and such properties are complete, still presuppose that there are persons as having those properties. According to Strawson, the properties like 'being at such and such time and place', having such and such weight and colour, and so on are M-predicates. The other properties are psychological properties like 'being in the state of happiness', or 'being in the state of pain', and so on are states of P-predicates. In this way, Strawson has rightly said, "the concept of a person is to be understood as the concept of type of entity such that both predicates ascribing corporeal characteristics, a physical situation and consciousness are equally applicable to an individual entities of that type." What is significant about them, as Strawson has pointed out is their co-applicability to the same person substance. The M-predicates cannot be ascribable independently because of that prohibits them from being ascribable to the conscious beings, like M-predicates, the P-predicates cannot be ascribed to the material bodies. This is because combination of a distinct kind of substance that has both physical and mental properties without being reducible to each other.

The above argument shows that Strawson accepting person as non-material and non-dual without rejecting Cartesian dualism. This is because Descartes held that when we are on the concept of person, we are really referring to one or both of two distinct substances of different types, each of which has its own appropriate types of states and properties. Each of which has its own appropriate types of states and properties, and none of the states belongs to both. That is to say that states of consciousness belong to one of these substances ant to the other. Descartes has given a sharp focus to this dualistic conception of person. It is not easy to get away from dualism because persons have both sorts of attributes such as mental and physical. According to dualistic conception, a person is something altogether distinct from body. That is, person is not identical with his body. Some dualists, however, believe that a person is a composite entity, one part of which is its body and another, part of which is something-immaterial spirit or soul. Thus dualism essentially adheres to the mind-body distinction and persons as mental beings as distinguished from material bodies.

According to Descartes, the self of a person is something altogether distinct from its body. So self is altogether non-physical, lacking in all physical characteristics whatever. On this interpretation, we can say that person is an immaterial substance a spirit or soul, which stands in some special relation to a certain physical body which is its body. Descartes thinks that a person is some sort of a combination of an immaterial soul and a physical body, which stand to one another in a rather mysterious relation of substantial union. However, Cartesian dualism does not maintain that a person is immaterial stuff. On the contrary, it maintains that a person is a combination of body and mind. In fact our bodies and we are utterly unlike one another in respect of the sorts of properties that we posses. Our bodies have spatial, extension, and a location in physical space, whereas we have no such qualities. On the other hand, we have thoughts and feelings, states of consciousness, whereas our bodies are known to have qualities other than these.

But the question arises, should a person not simply be identified with a certain physical body as Williams has argued? Strawson gives an answer to the above question. He says, mental states, such as thoughts and feelings seem not to be properly attributable to something like a body, but only to a person. One is inclined to urge that it is 'I' who thinks and feels, not my body, even it I need to have a body to be able to think and feel. However, if a person is composed if body but not identical with it, then it seems that every part of the body must be a part of the person but not every part of the person can be part of the body. So, one of the plausible assumptions is that a person has parts, which are not parts of his body, and so it is not identical with the body. However, by saying this we are denying that a person is composed of body. All that is meant is that persons have both bodily and mental existence. Persons are nor purely disembodied sprits.

A. J. Ayer says that the relation between consciousness and the subject to which is attributed is a contingent relation. According to him, a person is not a purely immaterial subject; it is rather an embodied person to which mental attributes are causally ascribable. He accepts a causal relation between the person and his body. Therefore, according to him, there is no contradiction in holding that a person's body would have been inhabited by another person. Strawson, however, rejects Ayers view, which takes person apart with the body. He rejects the idea of causal relation altogether. According to him, persons are more primitive than their mind and body. That is, persons are primary than whereas mind and body are secondary.

## Persons as Individuals

P.F. Strawson has adopted the term 'person' for a philosophical use which comes rather closer to common usage than did Locke's usage of the form, while it raises philosophical problems of its own perhaps it is less disreputable to hold that the person is a primitive concept. This is because the Lockian view of the concept of person is forensic concept, but Strawsonian concept of person is a metaphysical concept like the concept of the self, and therefore is not merely a social or a forensic concept. Pradhan pointed out that it is metaphysical precisely because it shows how it can be used to describe the minded being as the unique substance which is not identical with the body, though it is necessarily linked with the body. That is to say that persons have material bodies and yet are not on the same levels as the physical bodies or organisms. Persons, therefore, are not physical things at all and this is because persons transcend their physical existence.

The transcendental qualities, however, shows that persons are explainable from the first-person perspective. The first-person perspective are unique individual or an 'I' who experience, as Wittgenstein20 points out that even it is not 'name' which can substitute 'I'. Therefore, the first-person is not the descriptions of any human being, because it refers to third-person perspective, but it refers to the person himself or herself. This does not mean that person is distinct from this world, but person is part of this world. As Strawsonian person, to begin with, is to be understood as distinct from a mere material body, which retains the contrast customarily observed between person and things.

According to Strawson, each of us distinguishes between himself and the states of himself on the one hand, and what is not himself or a state of himself on the other. Then, the question is, what are the conditions of our making this distinction? In what way do we make it, and why do we make it in the way we do? Strawson argues that in our conceptual scheme, material bodies are basic particulars. This means that material bodies could be identified without reference to another individual in particular, whereas the identification and re-identification of particulars of other categories rest ultimately on the identification of material bodies.

Then, Strawson inquires whether we could make intelligible to ourselves a conceptual scheme in which material bodies are not basic. This leads him to the construction of a model no-space world, in which all the sensory items are auditory, but in which it did seem possible to find a place for the idea of a re-identifiable particulars by exploiting certain auditory analogues of the idea of spatial distance. However, the requirement was for a scheme in which a distinction was made between oneself and what is not one's self.

Let us now think of some ways in which we ordinarily talk of ourselves, certain things which we do, and which are ordinarily ascribed to ourselves. We ascribe to ourselves actions and intentions, sensations, thoughts and feelings, perceptions and memories. However, we ascribe to ourselves location and altitude. Of course, not only we ascribe ourselves temporary conditions, states, situations, but also enduring characteristics, including physical characteristics like height, shape and weight. That is to say, those among the things that we ascribe to ourselves are those that we also ascribe to material bodies. But there are things and attributes that we ascribe to ourselves, but cannot dream of ascribing to material bodies.

Let us take a visual experience. First, there is a group of empirical facts of which the most familiar is that if the eyes of that body are closed, the person sees nothing. To this group belong all the facts known to the ophthalmic surgeon. Secondly, there is the fact that what falls within his field of vision at any moment depends in part on the orientation of his eyes, that is, the direction his head is turned in and on the orientation of his eyeballs in their sockets. Thirdly, there is the fact that where he sees from or what his possible field of vision at any moment is, depends on his body.

Strawson divides these facts into three groups to emphasize the following - the fact that visual experiences, in all three ways, depend on facts about some body or bodies. It is a contingent fact that it is the same body. Each person's body occupies a special position in relation to that person's perceptual experience. For each person, there is one body occupies certain causal position in relation to that person's perceptual experience.

For Strawson, a person's body occupies an important position in the person's experience so that he could answer to the following questions satisfactorily:

a. Why are one's states of consciousness ascribed to anything at all? and

b. Why are they ascribed to the same thing as certain corporeal characteristics?

For the Cartesians this question does not arise, it is only a linguistic illusion that both kinds of predicate are properly ascribed to one and the same thing, and that there is a common owner or subject. Descartes says that when we speak of a person, we refer to two distinct substances. The state of consciousness belongs to one of these substances and not to the other. Strawson says that he escapes one of our questions, but it does not escape the other - why is one's state of consciousness ascribed at all, to anything?

In order to overcomes above problems Strawson used the concept of the person is "primitive concept." Then, he said, "What I mean by the concept of a person is the concept of a type of entity such that both predicates ascribing states of consciousness and predicates ascribing corporeal characteristics, a physical situation & c. are equally applicable to a single type." Now we can get answers to the above two question. Strawson said that answers to these two questions are connected in this manner, "... that a necessary condition of states of consciousness being ascribed at all is that they should be ascribed to the very same things as certain corporeal characteristics, a certain physical situation and c. That is to say, states of consciousness could not be ascribed at all, where as they were ascribed to persons, in the sense I have claimed for this world." Above Strawsonian view says that a necessary condition of a state of consciousness being ascribed at all is that they should be ascribed to persons. The concept of a person is prior to that of an individual consciousness. A person is not an embodied ego, but an ego might be a disembodied person.

Again, Strawson points out that one can ascribe states of consciousness to oneself only if one can ascribe them to others. One can ascribe them to others only if one can identity other subjects of experience. In addition, one cannot identify others if one can identify them only as subjects of experience, of states of consciousness. He says, this way will lead to Cartesianism. We cannot but refer to the bodies of others. So states of consciousness could not be ascribed at all, unless they are ascribed to an individual person who has a body. So the pure individual person or consciousness, in the sense of the pure ego, is a concept that cannot exist. Strawson says it can exist only as a secondary, non-primitive concept, and can be analyzed in terms of the concept of person.

The pure individual consciousness cannot exist as a primary concept to be used in the explanation of the concept of a person, but it might have a logically secondary existence. From within our conceptual scheme, each of us can conceive of his or her individual survival of bodily death. One has to think of oneself as having thoughts and memories in a disembodied state. But this disembodied state is only a secondary concept, because cannot but think of persons as embodied beings. According to Strawson, "A person is not an embodied ego, but an ego might be a disembodied person, retaining the logical benefit of individuality from having been a person." As we have seen, there are two kinds of predicates properly applied to individuals of this type. The first kind of predicates consist of those that are also properly applied to material bodies to which we do not ascribe states of consciousness, which he calls M-predicates. The second type consists of those predicates such as 'thinking hard', 'belief in God', etc., which he calls P-predicates. Therefore, Strawson says that the concept of person is to be understood as the concept of a type of entity such that both predicates ascribing states of consciousness and those ascribing corporeal characteristics (M-predicates) are equally applicable to an individual entity.

Then, he said, "... the concept of a person is to be understood as the concept of a type of entity such that both predicates ascribing states of consciousness and predicates ascribing corporeal characteristics, a physical situation and c. are equally applicable to an individual entity of that type." Strawson is not taking the concept of person as a secondary concept in relation to two primary kinds, that all, a particular consciousness and a particular body (human). Then Strawson says that " ... though not all P-predicates are what we should call 'predicates ascribing states of consciousness' (e.g. 'going for a walk' is not), they may be said to have this in common, that they imply the possession of consciousness on the part of that to which they are ascribed."

From the above stand point, what Strawson want to say is that "one ascribes P-predicates to others on the strength of observation of their behaviour; and that behavioural-criteria one goes on are not just signs of the presence of what is meant by the P-predicates, but are criteria one goes on are not just signs of the presence of what is meant by the P-predicates, but are criteria of a logically adequate kind for the ascription of the P-predicates." This claim shows that person is immaterial because of states of consciousness is applicable to person. This is because there are predicates, which could be both self-ascribable and other-ascribable to the same individual. But there remain many cases in which one has an entirely adequate basis for ascribing a P-predicate to oneself, and yet, this basis is distinct from those on which one ascribes the predicates to another. In other ward, these predicates have the same meaning and both ways of ascription is in perfect individual. That is why P-predicates have certain characteristics such as 'I am in pain', 'I am depressed', and etc., and one should not ascribe to some body from these observations because this leads to third-person perspective of the concept of person.

Moreover, the above explanation begs an important question. The question is: "How can one ascribe to oneself, not on the basis of observation, the very same thing that others may have, on the basis of observation, reason of a logically adequate kind for ascribing one, which might be phrased?" Strawson says that the above question may be observed in a wider one, which might be phrased. The questions are, " 'how are P-predicates possible?' or 'how is the concept of a person possible?'" Strawson says that these two questions are replacing those two earlier questions, that are, why are one's states of consciousness ascribed to anything at all? And why are they ascribed to the same thing as certain corporeal characteristics? The answer to these two questions are inherented in the primitiveness of the concept of person, this is because the uniqueness character of P-predicates. Because he or she, who is individual have the P-predicates. The attributes of P-predicates that make a person as an individual. The persons are metaphysical beings claim an ontological reality in the sense that they could not be what they are without a metaphysical essence.

## Creativity and Consciousness

Creativity is one of the aspects of intelligence and is one of the most important features of the human mind. It is creativity, in the very specific sense of the term used here, which distinguishes humans from machines. Now the question is: Under what conditions can we say that a human act is creative? We can identify two aspects in any act. One is the product of the act and the other is the process. By product, we mean that which is produced by the act. The process stands for the way the product is produced. The process, being psychological, is something subjective. Therefore, in order to judge whether an act is creative, it is not possible to depend only upon the features of the psychological processes involved. An act can be judged to be creative on the basis of some of the objective features that the product possesses, such as artistic creations, poetic compositions, etc. Therefore, the question, what is creativity comes down to: What are the characteristics features of a creative product in terms of which the act that produced it is judged to be creative?

### Features of Creativity

Firstly, one of the important features of creativity is novelty. By the term 'novelty', we mean that the product did not come into existence before the act in question was performed. The novelty of the creativity of a product lies in the fact that it is different from other products already existing in the same domain. We come to know this only after the object is produced. No prior knowledge of the antecedent events and processes or the circumstances that led to the production of object can help us to know in advance what features the product will have. As Vernon defined it, "creativity denotes a person's capacity to produce new or original ideas, insights, invention or artistic products, which are accepted by experts as being of scientific, aesthetic, social, or technical value."

In a similar manner Boden points out that "if we take seriously the dictionary-definition of creation, 'to bring into being or form out of nothing,' creativity seems to be not only beyond any scientific understanding, but even impossible. It is hardly surprising, then, that some people have 'explained' it in terms of divine inspiration, and many others in terms of some romantic intuition, or insight"[2] What Boden is trying to show is that if the creation is 'out of nothing', then it is God's creation because God alone can create something out of nothing. But we are here concerned with human creativity; this is because human creativity arises out of intuitions or out of the combinations of old ideas.

Once the product has come into existence, we may enumerate or list the features it possesses. But these features cannot be subsumed under a law or a rule. That is, statements describing the features of the object cannot be deduced from the rules or laws along with certain antecedent conditions. Thus "creativity is a puzzle, a paradox, some say a mystery. Inventors, scientists, and artists rarely know how their original ideas arise. They mention intuition, but cannot say how it works. Most psychologists cannot tell us much about it, either. What's more, many people assume that there will never be a scientific theory of creativity –for how could science possibly explain fundamental novelties? As if all this were not daunting enough, the apparent unpredictability of creativity seems to outlaw any systematic explanation, whether scientific or historical."[3] Thus Boden's definition of creativity brings out the features such as novelty, uniqueness and originality, which are essential to any creativity act. If a creative product has no value, no relevance, no originality, no novelty, and no uniqueness, then it is not new in its creation because there is nothing new in its creation. Whether a creation is out of something or out of nothing, these minimum features are essential to any creative act. Now the question is: Why should we be creative? We are creative because we have to solve our day-to-day problem. That is to say, we are creative in most of our day-to-day activities of problem solving. Hence, creativity is manifested in problem solving.

## Creativity as Problem Solving

We may understand creativity as problem solving. Thus a novel combination of ideas is said to be creative if it constitutes a solution to a problem. Problem solving is associated with many human activities. However, many questions arise such as: Are all problems well defined? Do we always know what the problem is? Are goals always clearly established? In many cases, the answer is no, so problem solving is not a mechanical affair, it is a creative act. Thus creative problem solving is different from the routine, mechanical ones.

According to Dodd and White, "problem solving, a frequent human activity, occurs when a goal cannot be achieved directly and a plan must be devised which will permit a goal attainment." On the other hand, Mayer defined it as, "problem solving is cognitive processing that is directed toward solving a problem." Here, the definition of problem solving consists of three components. Firstly, problem solving is cognitive act that occurs internally in the mind. Secondly, problem solving is a process having a definite direction and goal. That is why when a human being solves problem, he or she does a creative, insightful and intuitive act.

Moreover, when human beings solve problem, they identify the mental operations, representations, and strategies that they use when they solve problems. Dunbar proposes that problem solving consists of a search in a problem space, which has an initial state, a goal state, and a set of operations that can be applied in order to reach the goal. But everyone needs flexible, critical and creative thinking skills to cope with these problems and find solutions that can improve the physical and social environment. For creative problem solving, intelligence is necessary. An intelligent mind is a good thinker. Besides, a sense of humour helps in creative thinking because it relieves stress, tension, and monotony. It switches the mind into unexpected tasks. In order to solve problems, human beings should be creative, intelligent, and conscious. A conscious human being can solve the problem easily.

Though creativity is more likely to be observed among those who are more intellectually capable, such capability is not a guarantee of creativity? The ability assessed by IQ tests is not solely responsible for creative problem solving. Now the question is: What abilities distinguish creative from routine problem solving? Before attempting to identify the abilities responsible for creative problem solving, we must examine a model of intellectual functioning which distinguishes between forms of thought and the abilities underlying those forms. Dodd and White propose, after Guilford, " a tripartite division of intellectual functioning into contact, products and operations. The basic notion is that there are abilities associated with the processing of different forms of information; and that the ability applied to a particular task depends on the content of the task, the kind of output required of the problem solver, and the mental operations that must be performed on the content to produce a particular product."

Guilford's model is given below:

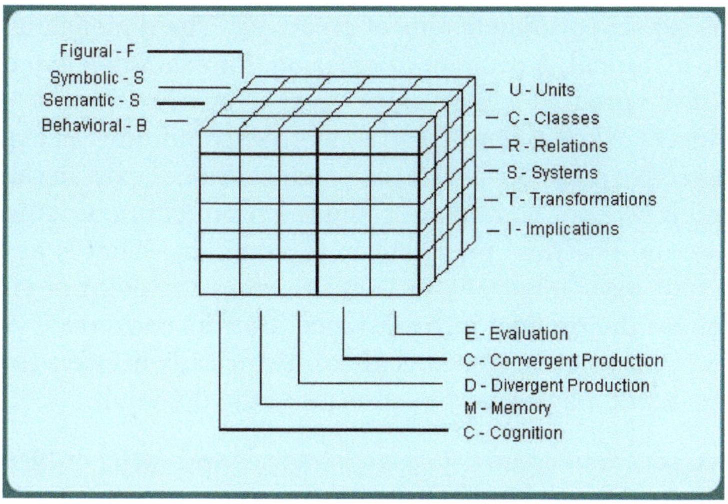


The figure shows five classes of mental operations to be performed on content. The cognition operations are those mental activities involved in representing the person's knowledge of the task. Memory refers to those mental activities that code and store information. Divergent and convergent production operations conducted on memory. The convergent thinker might say, 'What I do to solve this problem?' The divergent thinker might say, 'What are the ways of looking at this problem?'

Again, according to Gilford's model, there are four mental operations performed on various content. The first one is figural content. There are two figural content, visual-figural content and phonetic inputs. Semantic content consists of the imageless thoughts associated with percepts; an interpretation of the meaning of percepts constitutes semantic content. The semantic content is related to symbolic content symbols, and is used to communicate information. Finally, behavioural content refers to the nonverbal forms of communication that are received for processing. Product is a joint function of content and the mental operations that are applied to content. A unit can be anything having a definable identity. For example, any stimulus producing a visual or auditory percept, a word, memory, or behaviour can be a unit product. What is being claimed is that the intellect can produce as output anything that can be processed as input. Some units are similar and when combined form classes. If one unit suggests or implies another, an implication has been produced; there is a logical connection between the two elementary units.

Guilford's model plays a very important role in knowing how human brain solves problems. This model developes a theory of cognitive functioning that takes into a account creativity at various levels. And this model also facilitates research on creativity in domain of human intelligence. That is to say that this model shows how human beings solve problems.

## Dimensions of Creativity

There are various aspects or dimensions of creativity. The dimensions are psychological, and social or historical. A product or creation, for example, may be new in a psychological sense if the product is new to the creative agent. A product has special significance if the object strikes as new to the concerned community of experts. A product is new from an objective point of view, if the product did not exist in the domain before its production and it was not possible to bring the product into existence by following the available rules and practices prevalent in the domain. What is new objectively or socially must be new psychologically as well i.e., from the point of view of the agent whose action brought the product into existence. But the converse is not true: what is psychologically new may not be new socially or historically because the object considered as new by the agent may already be present in the domain.

Thus, we have two senses of creativity: psychological, and social or historical. The psychological sense is not divorced from the social sense of creativity, because as explained above, what is social is also psychological. By creativity in the social sense, we mean primarily the evaluation of the product as creative by a community of experts. As already noted, such evaluations are subject to socio-cultural factors and thus depend on many accidental factors. Since we cannot have a theory that deals with accidental factors responsible for the production and positive evaluation of the creative product, it is not possible to have a systematic explanation of creativity in the social sense. But, we can think of the psychological factors and processes involved in creativity, and underlying the historical aspects of creativity.

## Psychological Dimensions of Creativity

As discussed above, there are two senses of creativity i.e., psychological and historical. Boden characterizes them as 'P-creativity' and 'H-creativity' respectively. She writes, "A valuable idea is P-creative if the person in whose mind it arises could not have had it before; it does not matter how many times the other people have already had the same idea. By contrast, a valuable idea is H-creative if it is P-creative and no one else, in all human history, has ever had it before." According to her, it is not possible to have a 'theory that explains all and only H-creative ideas.' But in principle a psychological explanation of P-creative ideas is possible.

Now we examine the psychological sense of creativity within the framework of cognitive science. Cognitive science is a systematic study of the human cognitive capacities like thinking, perception, memory etc. The processes responsible for the exercise of these capacities are said to be internal to the system in question. The impacts of the socio-cultural and physical environment on these processes are not denied. But, it is assumed that the internal processes mediate such impacts. Therefore, cognitive science concentrates on a systematic study of the internal processes involved. The internal processes themselves are said to be a sort of computation and computation is understood as rule-governed symbol manipulation. Accordingly, if we are able to identify the symbol system and the rules that govern the transformation of the symbols, we may be in a position to account for the internal processes involved in cognition.

In the psychological sense, creative processes may be considered as internal cognitive processes that are very sophisticated in nature and may be understood as rule governed symbol-manipulation. So the key to cognitive modeling of creativity consists in identifying the symbol system involved and the rules that govern them. Since we are concerned with creativity in the psychological sense, we shall understand symbols as a system of ideas. Our attempts would be to understand how new ideas arise in the mind of the creative agent. One way of understanding it would be to conceive of new ideas as a result of the permutation and combination of old ideas. Through this process of permutation and combination, entirely unexpected, new and hitherto non-existent combinations of ideas emerge. However, this permutations and combinations of ideas are not random processes, rather they are rule-governed processes. However, all these combination of ideas must result in the generation of a new idea which were not already there. Then only these can be crative ideas.

However, all the novel ideas or thoughts by themselves would not mean that they are creative. We would consider the new combination of ideas to be somehow improbable and yet relevant. Boden suggests that there must be novelty in the creative ideas in the sense that the combination did not occur before. A creative idea, for her, is one that did not and could not have occurred before. Such ideas, according to Boden, are radically novel whereas ideas that did not but could have occurred before are merely novelties in a relative sense. In Boden's words, "many creative ideas, however, are surprising in a deeper way. They concern novel ideas that not only did not happen before, but that –in a sense to be clarified below –could not have happened before."

Furthermore, the key to understanding radical novelties lies in getting to know the meaning of 'could not' in this context. Boden writes, "Before considering just what this 'could not' means, we must distinguish two sense of creativity. One is psychological (let us call it P-creativity), the other historical (H-creativity). A valuable idea is P-creativity of the person in whose mind it arises could not have had it before; it does not matter how many times other people have already had the same idea. By contrast, a valuable idea is H-creative if it is P-creative and no one else, in all human history, has ever had it before." Boden clarifies it with the help of an example. Suppose a person comes up with an entirely new English sentence S, which has not been hitherto uttered by any one in the history of mankind. This sentence could have occurred before to a person who has internalized the grammar of English language and is familiar with its vocabulary. That is, the same sentence could have been produced by the same set of generative rules that produced other English sentences. In the same way, a new idea that could have been produced by the same set of generative rules that produced other familiar ideas is merely a first time novel idea. On the other hand, a radically novel idea or a creative idea is one that could not have been produced by the same system of generative rules that produces other familiar ideas. The above statements show that there are two kinds of creative thinking: divergent and convergent thinking. Thus the production of a genuinely original idea suggests that a specific and new generative system is available to the creative thinker. The generative system is not the product of random thinking but is a response to certain constraints on the kind of ideas that could be produced by the application of the generative system available to the creative agent before he came up with new generative rules. This shows that creativity is possible because of the constraints imposed by the available generative system of ideas. The existence of constraints demands that the creative agent come up with specific system of generative rules that permit radically novel combinations of ideas. This shows that the convergent creative thinking is a supplement to the divergent creative thinking, because in the case of divergent creative thinking it opens many aspects to have a creative idea. This shows that divergent thinking is opposed to convergent thinking. "Divergent thinking involves usual association of ideas, changing perspectives, and novel approaches to problem in contrast to convergent thinking, which involve linear logical steps."

## Historical Dimensions of Creativity

As we have already seen Boden has made a distinction between P-creativity and H-creativity. Historical dimension of creativity is opposed to psychological creativity because H creativity is new to the human history. As Boden remarks it, " . . . a valuable idea is H-creative if it is P-creative and no one else, in all human history has ever had it before." That is, H-creativity is typically associated with creativity in relation to the entire history of mankind. This type of creativity is not merely psychological but also social in character.

Furthermore, as Boden points out, " . . . there can be no systematic explanation of H-creativity, no theory that explains all and only H- creative ideas." What Boden is trying to show is that P–creativity or psychological creativity depends on H –creativity because by definition all H–creative is P-creative ideas, but not all P-creative ideas are H-creative. The psychological creativity (P-creative) is concerned with the individual psychology of the person concerned, where as H-creativity is a matter of social evaluation and collective judgment. Following this Brannigan writes, "Such value judgments are to some extent culturally relative, since what is valued by one person or social group may or may not be valued –praised, preserved, promoted by another."

H-creativity is opposed to P–creativity. In this sense, any H –creativity is 'more relative' than any merely P–creative idea. In the strict sense, we may not regard P –creativity as creative at all. In any case, P-creativity cannot be on par with H-creativity, because the latter alone guarantees novelty in all creative action.

## Are There Creative Machines?

Boden addresses the following questions regarding whether machines such as computers are creative. These questions are:

a. Can computers help us to understand human creativity?

b. Could computers do things which at least appear to be creative?

c. Could computers appear to recognize creativity?

d. Can computers really be creative?

According to Boden, the first question focuses on the creativity of human beings. The next two questions are psychological. The fourth question is a philosophical. Here, Boden is concerned with the first question, to which her answer is 'yes': computational concepts and theories can help us to specify the conceptual structures and processes in people's mind. In response to the fourth question, she says that computers can do things that appear to be creative, but whether we regard them as actually creative will depend on whether we are prepared to allow them 'a moral and intellectual respect comparable with the respect we feel for all human beings'? It is debatable whether machines can be ascribed the status of moral beings at all. Hence, Boden's response remains negative.

While Boden is concerned more with the way in which computers can help us to understand human creativity, Terry Dartnall is concerned with the fourth question more straightforwardly. Dartnall writes, "If machines cannot be creative then I doubt there is any significant sense in which they can be intelligent, for they will never 'have minds of their own'. I do mean this in the weak sense that they will always slavishly do what we tell them, but in the strong sense that they will never be able to generate their own ideas. And I take it as axiomatic that if they cannot generate their own ideas they cannot be intelligent."

Creativity is related to skills and abilities and also to ideas which are new and original. The ability to generate ideas and beliefs effectively *ex nihilo* is the core of creativity. The most common reason put forward to support the claim that computers cannot originate anything is that they merely follow instructions. The argument is:

"If X is merely following instructions, X is not being creative. Computers only follow instruction. Therefore, computers are not being creative."

In this argument, the first premise seems to be false, for we sometimes instruct people to be creative. For example, teacher advises the students to be creative and not mechanical. Therefore, it is possible to be creative and still be following instruction. But the fact is that computers are not like the students in this example. Computers merely follow instructions and cannot make a move on their own. Everything that a computer does is something that it was told to do. Hence, it cannot be said to be creative. The argument can be revised as follows:

"If everything that X does is something that it was told to do, then X is not creative. Everything that a computer does is something that it was told to do. Therefore, computers are not creative."

In this argument the second premise is false, if we do not instruct the computer in every action it they performs. If this premise were true, then we are required to give instruction at every step. But this may not be the case always. What Dartnall, however, means is that the machines do not literally follow the instructions but that the computer is built, or designed to respond in a predictable way to its instructions. So the argument can be further reformulated as follows:

"If X is designed to respond in predictable way to its instruction, then X is not creative. Computers are designed to respond in a predictable way to their instructions. Therefore, computers are not creative."

Still, this is not a strong argument, in view of the fact that creativity of computers cannot be denied just because they respond to the instructions of the designer. In this connection, one may appeal to Boden's distinction between P-creativity and H-creativity. Something is P-creative if it is fundamentally novel for the individual, and it is H-creative if it is fundamentally novel with respect to the whole of human history. The computers can be claimed to be P-creative if they can create something novel because they are not H–creative at all. But yet Dartnall argues, "There is, then, no obvious reason why they cannot have minds of their own. The final argument, that creativity is not predictable, is little more than a trick of the light." However, Drtnall's argument cannot prove that computers have creativity like human beings, since machine-creativity is a secondary phenomena in comparision to human creativity. The human P-creativity is a fundamental fact of human intelligence. The P-creativity of the human beings is supported and strengthened by the H-creativity.

## Creativity and Consciousness

This section deals with the problem of the relationship between creativity and consciousness. There are many philosophical problems which can be raised in this connection. They are: What is consciousness? What role, if any, does consciousness play in the explanation of creativity? Here, I am not arguing whether machines or robots have consciousness or creativity. I want to show how consciousness and creativity go together, and what role consciousness has in a creative act.

Philosophers have treated consciousness as a mystery for a long time. In recent years, researchers from diverse fields like psychology, neuroscience, computer science, physics, etc. are showing interest in the subject and are coming forward to share their findings with others. Consciousness is very much related to the creative activities because a human being cannot be creative without being conscious. This does not mean that a man who is conscious is necessarily creative, but consciousness is an essential feature of the human mind. But Antti Revonsuo rightly states, " . . . consciousness seems to form the center of our minds –it is the stuff that mental phenomena really are made of. What would be left of my mind, were all my conscious thoughts, beliefs, sensations, emotions and dreams eradicated? Without consciousness I would be a mere puppet, or a mindless zombie –a vacant body wandering around and going through human motions. Behind my eyes and the voice-patterns I utter there would be but darkness; there would be no subject there, which could in some way feel it exists."

If consciousness truly is such an essential feature of our mind, then the question: is it? Definitely, there are no universally accepted answers to this question. We are still in search of a true theory of consciousness. All we know about consciousness that it is a phenomena which cannot be measured, observed or experienced in public, because it is a subjective experience. It can be known only from a first-person perspective, but not from the third person/scientific/objective perspective. Because self is the subject, which feels, thinks and perceive. This is the qualitative character of human experience. As Thomas Nagel remarks, "an organism has conscious mental states if and only if there is something it is like to be that organism –something it is like for the organism. We may call this the subjective character of experience." All experience, therefore, are essentially subjective.

As Searle has argued, subjectivity is the most important feature of conscious mental states and processes. According to him, the conscious mental states do not have objective criteria and so are essentially first-person experiences. For him, subjectivity is an ontological category. Searle puts it, "But when we visualize the world with this inner eye, we cannot see consciousness. Indeed, it is the very subjectivity of consciousness that makes it visible in the crucial way. If we try to draw a picture of someone else's consciousness, we just end up drawing the other person (perhaps with a balloon growing out of his or her head). If we try to draw our own consciousness, we end up drawing whatever it is that we are conscious of." According to Searle, for first-person phenomenon of consciousness, is irreducible, and so cannot be explained objectively. It cannot be observed the way objective phenomena are observed. He comes to this conclusion by the following reasoning that the notion of observation, of seeing something, works on the presupposition that there is a distinction between the thing seen and the seeing of it. But for 'observation' there is simply no way to make this separation. Any introspection we have of our own conscious state is itself that conscious state. Consequently, "I cannot observe my own subjectivity, for any observation that I might care to make is itself that which was supposed to be observed."

Now we have to explore, what role, if any, does consciousness play in the explanation of creativity? In a general sense all conscious beings are creative, because creativity is a feature of consciousness. We human beings, manifest or show our creativity in our day-to-day life. For example, writing poems, musical compositions, scientific theories, painting, and many other things are creative acts. This also shows that creativity is an essential feature of mind or consciousness, because creativity presupposes that the creative being is conscious. As Boden says, "Human creativity is something of a mystery, not to say a paradox. One new idea may be creative, while another is merely new. What's the difference? And how is creatively possible? Creative ideas are unpredictable. Sometimes, they even seem to be impossible –and yet they happen . . ." There are many creative ideas, which have followed from human consciousness. For example, creating computers or robots is an unexpected use of everyday objects that could not have happened before. These machines have been created by creative minds. They themselves cannot be creative because they lack consciousness. Creativity is itself mystery which cannot be scientifically explained. As Boden writes, "Creativity is puzzle, a paradox, some say a mystery. Inventors, scientists, and artist rarely know how their original ideas arise. They maintain intuition, but cannot say how it works." Creativity is due to human 'intuition', which is beyond the scope of scientific investigation. Thus intuition is itself a creative process beyond the realm of scientific investigation.

## Nonreductive Mind

Qualia are the intrinsic quality of conscious experience. For example, the experience of tasting a sweet is very different from that of watching a movie because both of these have a different qualitative character of experience. This shows that there are different qualitative features of conscious experience. That is why; we cannot derive the pleasure of eating sweets by watching movies and via versa. As Chalmers writes, “a mental state is conscious if there is something it is like to be in the mental state. To put it in another way, we can say that a mental state is conscious if it has a qualitative feel–an associated quality of experience. These qualitative feels are also known as phenomenal qualities, or qualia for short.”[1] But, functionalists like Dennett have argued for eliminating qualia from the discourse of mind. The basic reason for them is that mind is a machine; it cannot entertain the so-called qualitative subjective experiences called the qualia. We have to show that the mentality of human mind cannot be represented in a mechanistic model and that there are subjective mental states which need a first-person explanation.

According to Dennett, “qualia are supposed to be properties of a subject that are (1) ineffable, (2) intrinsic, (3) private, (4) directly or immediately appraisable in consciousness.”[2] Qualia are ineffable because one cannot say exactly what way one is currently seeing, tasting, smelling, and so forth. Why qualia are ineffable is that they are intrinsic properties, which seems to imply inter alia that they are somehow atomic and unanaligible. Since they are simple, there is nothing to get hold of when trying to describe such property. Since qualia are ineffable and intrinsic, qualia are private because all interpersonal comparisons of these of appearing are systematically impossible. Lastly, since they are properties of experiences, qualia are directly accessible to the consciousness because qualia are properties of one's experiences with which one is immediately apprehensible in consciousness. Thus qualia constitute the phenomenal structure of the mind in that they enrich our understanding of the mind and also provide clues to the ontology of the mental. What the mental ultimately is, as distinguished from the physical, is to be known from what the qualia reveal about mind. Therefore, the qualia play a very important role in the understanding of mind.

The important question is: Is Dennett right in calling qualia the private and ineffable experiences of a queer sort? Obviously, not. As Pradhan argued, “the notion of privacy as we know from Wittgenstein's private language argument does not apply to the qualia in the sense that the qualia are intersubjectively intelligible and that they are available for inter-personal communication. The qualia of colour-perception are such that any two persons belonging to the same linguistic community can easily communicate their colour-experiences and can understand each other well. This shows that the qualia, in spite of being subjective, are not private at all. As to their effability or otherwise, it goes without saying that they are expressible in an interpersonal language; that is the reason why they are accessible to all speakers if they are suitably placed.”[3] Thus Dennett's main argument that the qualia are inaccessible to all except to the subject of the qualia does not hold good. Again, Dennett's argument that qualia are atomistic and non-relational is equally weak for the reason that the subjective experiences need not be atomistic at all because they can be taken as constituting the stream of consciousness in that they constitute a single unbroken series of the conscious experiences. In this sense the qualia are holistic rather than atomistic. The fact of the matter is that the qualia never exist in isolation and that they are always in a constellation. For example, the colour experience of a red rose is not only that of the colour red but also of the rose plant of certain shape and size. Here, the two experiences do not stand apart but constitute one whole.

Dennett is skeptical about the reality of the qualia because he believes qualia to be the private experiences and there is nothing in the mind that can correspond to these qualitative features of the mental states. According to him, the qualitative features are the appearances of the brain states, which in reality are the functional states of the brain. Dennett argues against qualia, because for him, the brain functions as a machine. The brain performs multiple functions; that is to say that all varieties of thought or all mental activities are accomplished in the brain by parallel, multitrack processes of interpretation and elaboration of sensory inputs. That is why this model of mind is called the multi-drafts-model. In Dennett's language, "According to the Multiple Drafts Model, all varieties of perception –indeed, all verities of thought or mental activity –are accomplished in the brain by parallel, multi-track processes of interpretation and elaboration of sensory inputs." The nature of the mind under this model is unfolded in the cognitive processes which the mind undertakes.

For Dennett, the mind turns out to be a computing machine programmed to cope with the cognitive representation of the world. For machine functionalists like him, the structure of the mind is the structure of the machine-representations. Therefore, in this respect, there is no place for the subjective qualia among the mechanical states of mind. Now the question is: Can the qualia be made a part of the third-person perspective? Dennett's reductionist program is fully committed to the reducibility of the qualia to the brain-state. However, this can be opposed on the ground that the qualia are ascribed to a conscious subject and not to the brain because the brain is a physical system though with infinite physical capacity. The subject is not reducible to the brain in the sense that brain itself belongs to the subject.

Our conscious mental states have different conscious experiences. For example, a man can see something as red today but tomorrow he may see the same thing as green. That is, the thing remaining same, a man's colour experience can vary from seeing red to seeing green. In this case, a person's colour experiences undergo an inversion in the sense that he sees something different from what he used to see earlier. Here, that man is not misidentifying the same object, rather he systematically goes on describing his previous experience of red as that of green now. Therefore, we cannot deny the logical possibility of our qualia being inverted in the case of oneself and of other. The qualia-inversion does not entail the physicalist and the machine-functionalist notion of consciousness because qualia inversion would not be possible if the conscious states would have been functional states of the brain. The qualia inversion cannot be ascribed to the physical and machine states. Therefore, the functionalist approach to consciousness must be rejected on the ground that consciousness states are not physical states because conscious states have qualia. As Shoemaker writes in the case of inverted spectrum, "there should be a systematic difference between the character of someone's colour experience at a certain time and the character of that same person's colour experience at another time." But it is conceivable that two people have similar functioning visual systems, but only the things look red to one-person while that looks green to the others. In this spectrum inversion, the way things look is possible but that cannot be given a functional description because person's mental life cannot be explicated in mechanical terms.

As we have mentioned earlier, there is a first-person dimension of the conscious states in that only from the first-person point of view we can understand the conscious states. The first-person point of view is such that it takes the mental states as belonging to a person from his or her subjective point of view. In this connection, we can mention Searle's view that the first-person perspective provides an ontological state to the subjective mental states. Searle writes, ". . . ontolological objectivity, is not an essential trait of science. If science is supposed to give an account of how the world works and if subjective states of consciousness are part of the world, then we should seek an (epistemically) objective account of an (ontologically) subjective reality, the reality of subjective states of consciousness. What I am arguing here is that we can have an epistemically objective science of a domain that is ontologically subjective." Mental states are subjective not in the epistemological sense of being known exclusively by the subject but in the ontological sense that they are essentially revealed only to subject. As Pradhan argues, "the mental life of man cannot be fully represented in a mechanistic system and that there are subjective mental states which need a first-person perspective for their proper understanding."

Whereas Putnam argues, functionalism is incompatible with our semantic externalism because functional organism is not simply a matter of 'sensory inputs', transition from one state to another, and 'motor outputs'. Semantic externalism refers to the content of our words and thoughts, which is partly determined by our relation with things in environment. A robot which has a program encoded into its system does not have any relation to the external environment. Putnam in his latter writings has rejected the computational view of mind on the ground that the literal Turing machine like the robot would not give a representation of the psychology of human beings and animals. For him, functionalism is wrong in holding the thesis that propositional attitude is just a computational state of the brain. For example, to believe that there is a cat on the mat, is not the same thing as that there is one physical state or a computational state believing that there is a cat on the mat. Then the question is whether these semantic and propositional attitudes properties and relations are 'reducible' to physical cum computational properties and relations. This is impossible because propositional attitudes refer to the intentional states, that is to say that it refers to various states of affairs in the world. For example, if I say that John will go to New Delhi from Hyderabad, this statement refers to many attitudes, and it cannot be realized computationally. Thus, according to Putnam, the functionalist is wrong in saying that semantic and propositional attitude predicates are semantically reducible to computational predicates.

There is no reason why the study of human cognition requires that we try to reduce cognition either to computations or to brain processes. We may well succeed in discovering theoretical models of the brain which vastly increase our understanding of how the brain works. But if we will reduce the human mind into brain, it no way helps us in understanding the mind. Therefore, functionalism fails to account for the real nature of the mental states because of its unsuccessful attempt to reduce mental states to the machine-states. It fails as a theory of mind because of its reductionist dogma. It makes mind meaningless in the universe. It also fails to explain how consciousness is possible. The mechanistic theory of mind does not have any positive or possible answer to the question how qualia are a necessary feature of consciousness. Artificial intelligence that offers a largely functionalist view of mind fails to explain how consciousness is possible.

We conclude that mechanistic explanation is not sufficient in explaining consciousness and creativity. This thesis follows from the conviction that we cannot conceive of consciousness unless we view it as having raw feelings. There are two aspects of this thesis, the epistemological and the metaphysical. Epistemologically, the subject of consciousness intimately knows the raw feelings. Metaphysically speaking, however, the raw feelings are real in the sense that they are part of the furniture of the mental world. Therefore, we can hardly deny that mental world is real. AI and cognitive science in general fails to recognize this fact about the mental world.

## Multiple Drafts Model

Daniel Dennett's multiple drafts model of consciousness is a physicalist theory of consciousness based upon cognitivism, which views the mind in terms of information processing. The theory is described in depth in his book, *Consciousness Explained*, published in 1991. As the title states, the book proposes a high-level explanation of consciousness which is consistent with support for the possibility of strong AI.

Dennett describes the theory as *first-person operationalism*. As he states it:

The Multiple Drafts model makes [the procedure of] "writing it down" in memory criterial for consciousness: that is *what it is* for the "given" to be "taken" ... There is no reality of conscious experience independent of the effects of various vehicles of content on subsequent action (and hence, of course, on memory).

## The Thesis of Multiple Drafts

Dennett's thesis is that our modern understanding of consciousness is unduly influenced by the ideas of René Descartes. To show why, he starts with a description of the phi illusion. In this experiment, two different coloured lights, with an angular separation of a few degrees at the eye, are flashed in succession. If the interval between the flashes is less than a second or so, the first light that is flashed appears to move across to the position of the second light. Furthermore, the light seems to change colour as it moves across the visual field. A green light will appear to turn red as it seems to move across to the position of a red light. Dennett asks how we could see the light change colour *before* the second light is observed.

Dennett claims that conventional explanations of the colour change boil down to either *Orwellian* or *Stalinesque* hypotheses, which he says are the result of Descartes' continued influence on our vision of the mind. In an Orwellian hypothesis, the subject comes to one conclusion, then goes back and changes that memory in light of subsequent events. This is akin to George Orwell's *Nineteen Eighty-Four*, where records of the past are routinely altered. In a Stalinesque hypothesis, the two events would be reconciled prior to entering the subject's consciousness, with the final result presented as fully resolved. This is akin to Joseph Stalin's show trials, where the verdict has been decided in advance and the trial is just a rote presentation.

[W]e can suppose, both theorists have *exactly* the same theory of what happens in your brain; they agree about just where and when in the brain the mistaken content enters the causal pathways; they just disagree about whether that location is to be deemed pre-experiential or post-experiential. [...] [T]hey even agree about how it ought to "feel" to subjects: Subjects should be unable to tell the difference between misbegotten experiences and immediately misremembered experiences. [p. 125, original emphasis.]

Dennett argues that there is no principled basis for picking one of these theories over the other, because they share a common error in supposing that there is a special time and place where unconscious processing becomes consciously experienced, entering into what Dennett calls the "Cartesian theatre". Both theories require us to cleanly divide a sequence of perceptions and reactions into before and after the instant that they reach the seat of consciousness, but he denies that there is any such moment, as it would lead to infinite regress. Instead, he asserts that there is no privileged place in the brain where consciousness happens. Dennett states that, "[t]here does not exist [...] a process such as 'recruitment of consciousness' (into what?), nor any place where the 'vehicle's arrival' is recognized (by whom?)."

Cartesian materialism is the view that there is a crucial finish line or boundary somewhere in the brain, marking a place where the order of arrival equals the order of "presentation" in experience because *what happens there* is what you are conscious of. ... Many theorists would insist that they have explicitly rejected such an obviously bad idea. But [...] the persuasive imagery of the Cartesian Theater keeps coming back to haunt us—laypeople and scientists alike—even after its ghostly dualism has been denounced and exorcized. [p. 107, original emphasis.]

With no theatre, there is no screen, hence no reason to re-present data after it has already been analysed. Dennett says that, "the Multiple Drafts model goes on to claim that the brain does not bother 'constructing' any representations that go to the trouble of 'filling in' the blanks. That would be a waste of time and (shall we say?) paint. The judgement is already in so we can get on with other tasks!"

According to the model, there are a variety of sensory inputs from a given event and also a variety of interpretations of these inputs. The sensory inputs arrive in the brain and are interpreted at different times, so a given event can give rise to a succession of discriminations, constituting the equivalent of multiple drafts of a story. As soon as each discrimination is accomplished, it becomes available for eliciting a behaviour; it does not have to wait to be presented at the theatre.

Like a number of other theories, the Multiple Drafts model understands conscious experience as taking time to occur, such that percepts do not instantaneously arise in the mind in their full richness. The distinction is that Dennett's theory denies any clear and unambiguous boundary separating conscious experiences from all other processing. According to Dennett, consciousness is to be found in the actions and flows of information from place to place, rather than some singular view containing our experience. There is no central experiencer who confers a durable stamp of approval on any particular draft.

Different parts of the neural processing assert more or less control at different times. For something to reach consciousness is akin to becoming famous, in that it must leave behind consequences by which it is remembered. To put it another way, consciousness is the property of having enough influence to affect what the mouth will say and the hands will do. Which inputs are "edited" into our drafts is not an exogenous act of supervision, but part of the self-organizing functioning of the network, and at the same level as the circuitry that conveys information bottom-up.

The conscious self is taken to exist as an abstraction visible at the level of the intentional stance, akin to a body of mass having a "centre of gravity". Analogously, Dennett refers to the self as the "centre of narrative gravity", a story we tell ourselves about our experiences. Consciousness exists, but not independently of behaviour and behavioural disposition, which can be studied through heterophenomenology.

The origin of this operationalist approach can be found in Dennett's immediately preceding work. Dennett (1988) explains consciousness in terms of *access consciousness* alone, denying the independent existence of what Ned Block has labeled *phenomenal consciousness*. He argues that "Everything real has properties, and since I don't deny the reality of conscious experience, I grant that conscious experience has properties". Having related all consciousness to properties, he concludes that they cannot be meaningfully distinguished from our judgements about them. He writes:

The infallibilist line on qualia treats them as properties of one's experience one cannot in principle misdiscover, and this is a mysterious doctrine (at least as mysterious as papal infallibility) unless we shift the emphasis a little and treat qualia as logical constructs out of subjects' qualia-judgments: a subject's experience has the quale F if and only if the subject judges his experience to have quale F. We can then treat such judgings as constitutive acts, in effect, bringing the quale into existence by the same sort of license as novelists have to determine the hair color of their characters by fiat. We do not ask how Dostoevski knows that Raskolnikov's hair is light brown.

In other words, once we've explained a perception fully in terms of how it affects us, there is nothing left to explain. In particular, there is no such thing as a perception which may be considered in and of itself (a quale). Instead, the subject's honest reports of how things seem to them are inherently authoritative on how things seem to them, but not on the matter of how things actually are.

So when we look one last time at our original characterization of qualia, as ineffable, intrinsic, private, directly apprehensible properties of experience, we find that there is nothing to fill the bill. In their place are relatively or practically ineffable public properties we can refer to indirectly via reference to our private property-detectors — private only in the sense of idiosyncratic. And insofar as we wish to cling to our subjective authority about the occurrence within us of states of certain types or with certain properties, we can have some authority — not infallibility or incorrigibility, but something better than sheer guessing — but only if we restrict ourselves to relational, extrinsic properties like the power of certain internal states of ours to provoke acts of apparent re-identification. So contrary to what seems obvious at first blush, there simply are no qualia at all.

The key to the multiple drafts model is that, after removing qualia, explaining consciousness boils down to explaining the behaviour we recognise as conscious. Consciousness is as consciousness does.

## Critical Responses to Multiple Drafts

Some of the criticism of Dennett's theory is due to the perceived tone of his presentation. As one grudging supporter admits, "there is much in this book that is disputable. And Dennett is at times aggravatingly smug and confident about the merits of his arguments [...] All in all Dennett's book is annoying, frustrating, insightful, provocative and above all annoying." (Korb 1993)

Bogen (1992) points out that the brain is bilaterally symmetrical. That being the case, if Cartesian materialism is true, there might be *two* Cartesian theatres, so arguments against only one are flawed. Velmans (1992) argues that the phi effect and the cutaneous rabbit illusion demonstrate that there is a delay whilst modelling occurs and that this delay was discovered by Libet.

It has also been claimed that the argument in the multiple drafts model does not support its conclusion.

## "Straw Man"

Much of the criticism asserts that Dennett's theory attacks the wrong target, failing to explain what it claims to. Chalmers (1996) maintains that Dennett has produced no more than a theory of how subjects report events. Some even parody the title of the book as "Consciousness Explained Away", accusing him of greedy reductionism. Another line of criticism disputes the accuracy of Dennett's characterisations of existing theories:

The now standard response to Dennett's project is that he has picked a fight with a straw man. Cartesian materialism, it is alleged, is an impossibly naive account of phenomenal consciousness held by no one currently working in cognitive science or the philosophy of mind. Consequently, whatever the effectiveness of Dennett's demolition job, it is fundamentally misdirected.

## Unoriginality

Multiple drafts is also attacked for making a claim to novelty. It may be the case, however, that such attacks mistake which features Dennett is claiming as novel. Korb states that, "I believe that the central thesis will be relatively uncontentious for most cognitive scientists, but that its use as a cleaning solvent for messy puzzles will be viewed less happily in most quarters." (Korb 1993) In this way, Dennett uses uncontroversial ideas towards more controversial ends, leaving him open to claims of unoriginality when uncontroversial parts are focused upon.

Even the notion of consciousness as drafts is not unique to Dennett. According to Hankins, Dieter Teichert suggests that Paul Ricoeur's theories agree with Dennett's on the notion that "the self is basically a narrative entity, and that any attempt to give it a free-floating independent status is misguided." [Hankins] Others see Derrida's (1982) representationalism as consistent with the notion of a mind that has perceptually changing content without a definitive present instant.

To those who believe that consciousness entails something more than behaving in all ways conscious, Dennett's view is seen as eliminativist, since it denies the existence of qualia and the possibility of philosophical zombies. However, Dennett is not denying the existence of the mind or of consciousness, only what he considers a naive view of them. The point of contention is whether Dennett's own definitions are indeed more accurate, whether what we think of when we speak of perceptions and consciousness can be understood in terms of nothing more than their effect on behaviour.

## Information Processing and Consciousness

The role of information processing in consciousness has been criticised by John Searle who, in his Chinese room argument, states that he cannot find anything that could be recognised as conscious experience in a system that relies solely on motions of things from place to place. Dennett sees this argument as misleading, arguing that consciousness is not to be found in a specific part of the system, but in the actions of the whole. In essence, he denies that consciousness requires something in addition to capacity for behaviour, saying that philosophers such as Searle, "just can't imagine how understanding could be a property that emerges from lots of distributed quasi-understanding in a large system".

## Inverted Spectrum

Inverted spectrum is the hypothetical concept of two people sharing their color vocabulary and discriminations, although the colors one sees — one's qualia — are systematically different from the colors the other person sees.

Inverted qualia

The concept dates back to John Locke. It invites us to imagine that we wake up one morning, and find that for some unknown reason all the colors in the world have been inverted. Furthermore, we discover that no physical changes have occurred in our brains or bodies that would explain this phenomenon. Supporters of the hypothesis of qualia as non-physical entities argue that, since we can imagine this happening without contradiction, it follows that we are imagining a change in a property that determines the way things look to us, but that has no physical basis. In more detail:

1. Metaphysical identity holds of necessity
2. If something is possibly false, it is not necessary
3. It is conceivable that qualia could have a different relationship to physical brain-states
4. If it is conceivable, then it is possible
5. Since it is possible for qualia to have a different relationship with physical brain-states, they cannot be identical to brain states (by 1).
6. Therefore, qualia are non-physical.

The argument thus states that if we find the inverted spectrum plausible, we must admit that qualia exist (and are non-physical). Some philosophers find it absurd that an "armchair argument" can prove something to exist, and the detailed argument does involve many assumptions about conceivability and possibility, which are open to criticism. Perhaps it is not possible for a given brain state to produce anything other than a given quale in our universe, and that is all that matters.

The idea that an inverted spectrum would be undetectable is also open to criticism on more scientific grounds:

> there are more perceptually distinguishable shades between red and blue than there are between green and yellow, which would make red-green inversion behaviorally detectable. And there are yet further asymmetries. Dark yellow is brown (qualitatively different from yellow), whereas dark blue is blue[..] Similarly, desaturated bluish-red is pink (qualitatively different from saturated bluish-red), whereas desaturated greenish-yellow is similar to saturated greenish-yellow. Again, red is a "warm" color, whereas blue is "cool" — and perhaps this is not a matter of learned associations with temperature.

Inverted spectrum arguments have applications to behavioralism, physicalism, representationalism, functionalism, skepticism and the hard problem of consciousness.

In his book *I Am A Strange Loop,* Douglas Hofstadter argues that the inverted spectrum argument entails a form of solipsism in which people can have no idea about what goes on in the minds of others—contrary to the central theme of his work. He presents several variants to demonstrate the absurdity of this idea: the "inverted political spectrum", in which one person's concept of liberty is identical to another's concept of imprisonment; an inverted "sonic spectrum" in which low musical notes sound like "high" ones and vice versa (which he says is impossible because low sounds can be felt physically as vibrations); and a version in which random, complex qualia such as riding a roller coaster or opening presents are reversed, so that everyone perceives the world in radically different, unknowable ways.

## First-Person

The first-person point of view associates the phenomenon of consciousness with the conscious subject. That is to say, consciousness is grounded on the very nature of a conscious being. So being conscious has to be understood from the subjective point of view of a conscious being. According to the Cartesian conception, we have access to the contents of our own minds in a way denied to us in respect of matter. That is why, we can know what we think, feel and want with a special kind of certainty. There is something special about our knowledge of our own minds that naturally goes with the Cartesian view. In other words, a subject (or self) has first person authority with respect to the contents of his or her mind, whereas others (third persons) can only get at these contents indirectly. We count as an authority about our own minds, because we can know about it directly or immediately and this is very different from others' knowledge.

According to the first-person account, we ourselves are conscious and have direct access to our consciousness. That is, we can find out what consciousness is from our own case. When we wake-up in the morning from a deep, dreamless sleep, we can notice the presence of a state qualitatively different from what happened before. In some cases like having an after image, making a decision, etc., we know all our conscious states from our own experience.

In the first-person account, there is possibility of a private ostensive definition. We can explain the meaning of an expression to someone by giving him/her a series of examples of the things to which the expression refers. For example, to explain to someone what the colour-words like 'red', 'green', 'yellow', etc., are, we would give him/her some examples of what red and green things are. According to some philosophers, the first-person account must have at least some sensation words by ostensive definition, by being presented with examples of particular sensations. Every person must have the sensation himself/herself. A person, who has never experienced pain, (in the case of abnormal nervous people, who are incapable of feeling pain) would not fully understand the expressions like 'pain', 'ache', or 'twinge' on this account. The person himself must experience the sensation to learn the full meaning of the word sensation. The painting to one's own experience is a private ostensive definition. On the first-person account, this element of 'privacy' is an essential part of the meanings of sensation words. If a person 'x' feels pain, only he can have that feeling, others may feel their own pain, but they cannot feel the pain of 'x'. They may be able to tell from the situation and his behaviour that he is in pain, but they cannot tell what his feeling of pain is. His pain is private to himself; no one else can feel it. Anything any other person feels will be that of his or her own feelings.

There are two difficulties that can arise in this connection. First, there is the problem of verification: i.e., could we never have any grounds for applying them in case of others? After all, we can never observe another's inner sensation; we can do that only in our own case. We all can observe the outword behaviour as far as others are concerned. So, could we say to another person that he has the inner sensation? This problem is the problem of other minds. That means, how can one person believe that the inner states of what he/she knows as consciousness in his/her own case ever occurs in other person's case other than his/her own case? Second, there is the problem of meaning. How could such expressions, which get their meanings from what he/she experiences in his/her own case, apply to others?

The third-person perspective presents an objective picture of consciousness purely from an impersonal point of view. That is why, it cannot present the raw feel of experience. The first person perspective gives us more interesting information than third person point of view. It is natural to claim that the first-person ascriptions are grounded on the mental states themselves or they are related with mental states themselves. Shoemaker writes, "One can, apparently, have this knowledge without presuming anything about the connections between the mental states and the bodily states of affairs, behavioural or physiological, which serve as the evidence for our ascription of these same mental states to other persons. Thus the first-person perspective apparently gives one a freer rein than the third-person perspective in investigating, empirically, the connections between mental states of affairs and bodily ones. And so if we are concerned with what the possibilities are, with respect to these connections, imagining what we could discover from the first-person perspective seems potentially more revealing than imagining what we could discover from the third-person perspective."

From the first-person perspective, the brain-mind identity theory does not hold good. One can imagine that someone is in pain without there being any C-fiber stimulation going on. Here Shoemaker argues that it is easy enough to imagine a case in which C-fiber stimulation is going on in someone and that person is not in pain. But the question is whether it is true that every brain state can be imagined to occur without the subject being in pain. He argues, "We cannot be in a position to judge about someone *both* that she is not in pain *and* that she is in a state that influences her behaviour in just the ways we think pain influences behaviour...someone can be in pain when there is no behavioural evidence that she is, and when there is behavioural evidence that she is not – she may be successfully suppressing the manifestations of pain. And in such a case we will normally believe, mistakenly, that the person is not in pain...the person's brain is in the state C-fiber-stimulation-plus-and if we don't realize that this is an optimal candidate for being a total realization of pain, we may continue to believe that the person is not in pain while believing that she has in her brain what is in fact an optimal candidate for being a total realization of pain."

In this connection, Searle discusses the example of silicon chips thought experiment in which the parts of someone's brain are progressively replaced by silicon chips. Here the procedure starts with a treatment for blindness due to deterioration of the brain and the doctor replaces silicon chips into the visual cortex. According to him, what the thought- experiment shows " ...we imagine that your brain is entirely replaced by silicon chips; that as you shake your head, you can hear the chips rattling around inside your skull. In such a situation there would be various possibilities...you continue to have all of the sorts of thoughts, experiences, memories, etc., that you had previously; the sequence of your mental life remains unaffected. In this case, we are imagining that the silicon chips have the power not only to duplicate your input-output functions, but also to duplicate the mental phenomena, conscious and otherwise, that are normally responsible for your input-output functions."

Searle further writes, "...as the silicon is progressively implanted into your dwindling brain, you find that the area of your conscious experience is shrinking, but that this shows no effect on your external behaviour. You find, to your total amazement, that you are indeed losing control of your external behaviour. You find, for example, that when the doctors test your vision, you hear them say, "We are holding up a red object in front of you; please tell us what you see." You want to cry out, "I can't see anything. I'm going totally blind." But you hear your voice saying in a way that is completely out of your control, "I see a red object in front of me.""

In these thought experiments the first-person experiences are missing. The external observable behaviour remains the same, but conscious experience gradually shrinks to nothing. From the outside, it observes that the person is fine, but from the inside, he is gradually dying. Here, a certain physical makeup fails to support mentality and consciousness. Searle says, "we are imagining a situation where you are eventually mentally dead, where you have no conscious mental life whatever, but your externally observable behaviour remains the same."

In above thought experiments, Searle shows the causal relationships between brain processes, mental processes, and externally observable behaviour. The silicon chips are able to duplicate certain input-output functions of the brain; they also maintain heart-beat, respiration, and other vital processes but the patient's conscious mental life is left out. Here the ontology of the mental is essentially a first-person ontology. Mental states only exist as subjective, first-person phenomena. The first-person point of view is quite different from the third-person point of view, because, from the third-person point of view, one cannot be able to tell whether one had mental states at all.

To explain experience, a first-person perspective is needed. A common conception of how we experience the outer world is that something that is a thought or information is transferred from the outer world into the sphere of the first- person. There is something that is the mind, and, therefore, there is something that it is like to be that mind-and it is a first-person experience. If there is anything to which we have a distinctive first-person access, it is the "phenomenal" aspects of such states to which Nagel called "what it is like" to have them. If one views a bat from the "third-person" external observer's perspective, one may be able to observe everything that, in principle, there is to observe about bat physiology and behaviour. But one cannot observe how the bat experiences the world. The point is how broad one's knowledge may be of bat physiology and behaviour, one cannot know what it is like to be bat from the bat's point of view. Therefore, for Nagel, bat consciousness could never be just a construct within an information processing model.

The first-person approach explains consciousness as a metaphysical problem, whereas the third-person approach treats it as a scientific problem. In third-person approach, consciousness is explained as a problem of science like heat, life or nuclear physics. In general, this approach has much success as a study of mental processes in cognitive science. In other words, it discusses mind as a cognitive system. The first-person approach describes what it is like to be, who we are, and what it might be like to be something different from another human, a bat, or a computational system. It is these subjective questions, which pose the real problem of consciousness.

We can divide the first-person problem into three parts such as the problem of sensory qualia, subjective mental content, and the existence of subjective experience. In case of sensory qualia, these are the qualitative aspects of our mental states, of our sensations. From the third–person approach, the nature of qualia cannot be explained. In the case of second problem, the subjective mental content is not entirely different from that of sensory qualia i.e., the experience of content is itself qualitative. For example, when we think about a tree, there is something that takes place in our subjective experience that has something to do with the tree. The third problem is the existence of subjective experience. The problem is why subjective states should exist in the first place. Many philosophers believe in zombies – humans with normal behaviour but without any subjective mental states. These may be logically possible, but it seems implausible that there could be such things in the actual world. However, for every first-person (conscious) mental sate, there is a corresponding third-person (functionally) mental state. For every subjective sensation, there corresponds an objectively characterisable perception.

Consciousness, the subjective experience of qualia and mental content cannot be explained by a third-person account. Neuroscience and cognitive modeling do not explain the qualitative nature of a sensation of red, or even why such a subjective sensation should exist. From the first-person perspective, we have beliefs about our own experiences and attitudes and we have phenomenal consciousness, which is distinguished from other features. Many philosophers agree that consciousness is a surprising phenomenon. It is a brute fact about us that we have first-person experiences.

There is a sense in which each person's consciousness is subjective to that person, a sense in which he is related to his pains, tickles, itches, thoughts and feelings in a way that it is quite unlike that others are related to these qualities. This phenomenon can be described in various ways. It is sometimes described as that feature of consciousness because of which one can say that there is something that it is like or something that it feels like to be in a certain conscious state. That is to say that conscious states have a certain qualitative character, which are sometimes described as `qualia'. This subjective character of experience is not explainable by any functional or causal analysis, which is called by Chalmers the hard problem for consciousness.

## The 'Hard' Problem of Consciousness

The hard problem of consciousness, as Chalmers has shown, is the problem of experience, especially to first-person character which cannot be explained within a scientific framework. Cognitive science can explain a system's functions in terms of its internal mechanism. But it is not possible to explain what it is to have subjective experiences, because it is not a problem about the performance of functions. As Nagel argues, "Conscious experience is wide spread phenomenon. . . . fundamentally an organism has conscious mental states if and only if there is something that it is like to be that organism –something it is like for the organism." In recent times, all sorts of mental phenomena have yielded to scientific explanation, but consciousness has stubbornly resisted this explanation. Many philosophers and scientists have tried to explain it, but the explanations always seem to fall short of the target. Now the question is: why is it so difficult to explain? According to Chalmers, cognitive science has not explained, why there is conscious experience at all. When we think and perceive, there is a whir of information processing, but there are also subjective individual aspects of consciousness, which go beyond the information processing.

Chalmers writes, "When it comes to conscious experience, this sort of explanation fails. What makes the hard problem hard and almost unique is that it goes beyond problems about the performance of functions. To see this, not that even when we have explained the performance of all the cognitive and behavioural functions, in the vicinity of experience –perceptual discrimination, categorization, internal access, verbal report –there may still remain a further question: Why is the performance of these functions accompanied by experience?" According to him, even if all the functions of a system are well articulated, there is further question as to why there is any experience at all accompanying their function. Cognitive science fails to explain why there is any experience at all, even though it explains all the brain functions.

As Chalmers argues that the hard problem of consciousness consists in the 'why' questions regarding consciousness. But the question is: why is the 'hard' problem so hard? And why are the easy problems so easy? The easy problems are easy because it concern the explanation of cognitive abilities and functions. To explain a cognitive function, we need a mechanism that can perform the function. The cognitive sciences offer this type of explanation and so are well suited to the easy problem of consciousness. On the other hand, the 'hard' problem is 'hard', because it is not a problem about the performance of functions. The problem persists even when the performance of all the relevant functions are explained. Chalmers says, "I suggest that a theory of consciousness should take experience as fundamental. We know that a theory of consciousness requires the addition of something fundamental to our ontology, as everything in physical theory is compatible with the absence of consciousness. We might add some entirely new non-physical feature, from which experience can be derived, but it is 'hard' to see what such a feature would be like. More likely, we will take experience itself as a fundamental feature of the world, alongside mass, charge, and space-time. If we take experience as fundamental, then we can go about the business of constructing a theory of experience."

Machine Intelligence has not solved the 'hard' problem of consciousness because, as we have seen, it has explained consciousness only in terms of the easy problem of consciousness. Easy problems are all concerned with how a cognitive or behavioural function is performed. These are questions about how the brain carries out the cognitive task that is, how it discriminates stimuli, integrates information and so on. Whereas the hard problem of consciousness goes beyond the problems about how functions are performed. If artificial intelligence tries to give a definite definition of consciousness then it leaves out the explanatory gap, that is to say, it discusses the distinction between mind and body. If this is so, then it leaves out subjective experience, and opts for a third-person perspective of consciousness.

## The Explanatory Gap and Subjectivity

Consciousness makes the mind-body problem really intractable. The reductionists deny that there is a mind-body problem at all. For them, there is no explanatory gap between mind and body. Because there is no distinction between mind and body. Mind can be explained in terms of body, and there is nothing called the mind, since the mind itself is a part of the body. Therefore, for them, the mind is reductively explainable in terms of body. On the other hand, many philosophers hold that mental states are not reducible to any physical state(s). That is, the mental states are not reductively explainable. That is to say that that no reductive explanation of consciousness can succeed, because there is subjective quality of experience. Therefore, he argues that this quality of consciousness makes it different from all other properties, including emergent biological properties such as life.

The essence of body is spatial extension, the essence of mind is thought. Thought is taken to be the defining attribute of mind which is an incorporeal substance a substance that is non-spatial in nature. Chalmers writes, "By the term 'thought,' I understand everything which we are aware of as happening within us, in so far as we have awareness of it." What follows from Descartes's view are that consciousness is essentially a first-person, subjective phenomena, and conscious states cannot be reduced or eliminated into third-person. Therefore, it is consciousness, which makes the explanatory gap between the first person and third-person perspective. According to the Cartesian conception, we have access to the contents of our own minds in a way denied to us in respect to matter. There is something special about our own knowledge of our own minds that naturally goes with the Cartesian view.

However, the mental life with its qualia cannot be nomologically determined by the physical conditions of the universe. The following are the reasons for the thesis that the mental life is independent of the physical body, though they co-exist: "(A). The qualia of the mental states cannot be reproduced in an artificial machine like a robot or a machines table; they are unique to the person concerned. (B). The qualia are the essence of consciousness and so must be intrinsic to the conscious subjects." Thus Pradhan concludes that the intelligibility gap between the qualia and the physical world remains, as the qualia are understood widely as belonging to the conscious subjects.

Consciousness, according to Nagel, makes the gap between mind and body, and 'subjectivity' is its most troublesome feature. Self is the subject, which encompasses our feelings, thinking, and perception. The qualitative character of experience is what it is like for it's subject to have the experience. As he puts it, "Conscious experience is a widespread phenomenon. It occurs at many levels of animal life, though we cannot be sure of its presence in the simpler organisms, and it is very difficult to say in general what provides evidence of it... no matter how the form may vary, the fact that an organism has conscious experience *at all* means, basically, that there is something it is like to *be* that organism...But fundamentally an organism has conscious mental states if and only if there is something it is like to *be* that organism—something it is like *for* the organism."

As we have seen, subjectivity cannot be explained reductively. Again, as Nagel argues, "It is not analyzable in terms of any explanatory system of functional states, or intentional states, since they could be ascribed to robots or automata that behaved like people though they experienced nothing." There is a subjective feeling attached to our conscious experience because subjective feelings are the outcome of our conscious experience. That is, consciousness itself cannot be established simply on the basis of what we observe about the brain and its physical effects. We cannot explain which property of the brain accounts for consciousness. Distinct cognitive properties, namely perception and introspection, necessarily mediate our relationships with the brain and with consciousness. We cannot understand how the subjective aspects of experience depend upon the brain that is really the problem.

Consciousness, according to Searle, is essentially subjective. This is not a mechanical state, as many philosophers believe. Some of these biological systems are conscious and that consciousness is essentially subjective. The term 'pain' is subjective as it is not accessible to any observer, because it is a first person experience. For example, I have a pain in my leg. In this case, the statement is completely subjective. The pain itself has a subjective mode of existence. As Searle puts it, "Conscious states exist only when they are experienced by some human or animal subject. In that sense, they are essentially subjective. I used to treat subjectivity and qualitativeness as distinct features, but it now seems to me that properly understood, qualitativeness implies subjectivity, because in order for there to be a qualitative feel to some event, there must be some subject that experiences the event. No subjectivity, no experience."

That is to say that the qualitative experience can exist only as experienced by some subjects. Because conscious states are subjective in this sense, it is legitimate to hold that there is first-person ontology, as opposed to the third-person ontology of mountains and molecules, which can exist even when there are no living creatures. Therefore, subjective conscious states have first-person ontology because they exist only when they are experienced by a subject as self. It is 'I' who has experience and in this sense, it has the subjective existence. This gap between the self and the body not only establishes explanatory gap, but also gives the ontology of first-person. Therefore, the 'subjectivity or 'I' is the central problem of the explanatory gap. Cognitive science tries to explain how conscious experience arises from the electrical process of the brain. But it cannot show how and why conscious states belong to the subject or I. This qualitative feature of mental states brings is the existence of qualia, which are the qualitative experiences of the human mind.

# Permissions

All chapters in this book are published with permission under the Creative Commons Attribution Share Alike License or equivalent. Every chapter published in this book has been scrutinized by our experts. Their significance has been extensively debated. The topics covered herein carry significant information for a comprehensive understanding. They may even be implemented as practical applications or may be referred to as a beginning point for further studies.

We would like to thank the editorial team for lending their expertise to make the book truly unique. They have played a crucial role in the development of this book. Without their invaluable contributions this book wouldn't have been possible. They have made vital efforts to compile up to date information on the varied aspects of this subject to make this book a valuable addition to the collection of many professionals and students.

This book was conceptualized with the vision of imparting up-to-date and integrated information in this field. To ensure the same, a matchless editorial board was set up. Every individual on the board went through rigorous rounds of assessment to prove their worth. After which they invested a large part of their time researching and compiling the most relevant data for our readers.

The editorial board has been involved in producing this book since its inception. They have spent rigorous hours researching and exploring the diverse topics which have resulted in the successful publishing of this book. They have passed on their knowledge of decades through this book. To expedite this challenging task, the publisher supported the team at every step. A small team of assistant editors was also appointed to further simplify the editing procedure and attain best results for the readers.

Apart from the editorial board, the designing team has also invested a significant amount of their time in understanding the subject and creating the most relevant covers. They scrutinized every image to scout for the most suitable representation of the subject and create an appropriate cover for the book.

The publishing team has been an ardent support to the editorial, designing and production team. Their endless efforts to recruit the best for this project, has resulted in the accomplishment of this book. They are a veteran in the field of academics and their pool of knowledge is as vast as their experience in printing. Their expertise and guidance has proved useful at every step. Their uncompromising quality standards have made this book an exceptional effort. Their encouragement from time to time has been an inspiration for everyone.

The publisher and the editorial board hope that this book will prove to be a valuable piece of knowledge for students, practitioners and scholars across the globe.

# Index